Children Are Mathematical Problem Solvers

By

Lynae Sakshaug
SUNY Brockport
Brockport, New York

Melfried Olson
Western Illinois University
Macomb, Illinois

Judith Olson
Western Illinois University
Macomb, Illinois

Copyright © 2002 by
THE NATIONAL COUNCIL OF TEACHERS OF MATHEMATICS, INC.
1906 Association Drive, Reston, VA 20191-1502
(703) 620-9840; (800) 235-7566; www.nctm.org

All rights reserved

Third printing 2006

ISBN 0–87353–529–4

The National Council of Teachers of Mathematics is a public voice of mathematics education, providing vision, leadership, and professional development to support teachers in ensuring mathematics learning of the highest quality for all students.

For permission to photocopy or use material electronically from *Children Are Mathematical Problem Solvers*, ISBN 87353-529-4, please access www.copyright.com or contact the Copyright Clearance Center, Inc. (CCC), 222 Rosewood Drive, Danvers, MA 01923, 978-750-8400. CCC is a not-for-profit organization that provides licenses and registration for a variety of users. Permission does not automatically extend to any items identified as reprinted by permission of other publishers and copyright holders. Such items must be excluded unless separate permissions are obtained. It will be the responsibility of the user to identify such materials and obtain the permissions.

The publications of the National Council of Teachers of Mathematics present a variety of viewpoints. The views expressed or implied in this publication, unless otherwise noted, should not be interpreted as official positions of the Council.

Printed in the United States of America

CONTENTS

Acknowledgment iv
Introduction. v
An Overview of Problems and Problem Solving vi

1	Animals and Fences at a Zoo	1
2	Code It Be?	5
3	How Many Rectangles?	10
4	Double That Dough	14
5	How Much Film?	17
6	When Will We Reach 1/2?	21
7	Decoration Delight	24
8	You Gotta Have Heart and Blood!	28
9	How Many Sandwiches	32
10	Shuffling a Line	36
11	Counting Squares	39
12	How Many Times Can You Take 1/2?	42
13	Guess the Weight	45
14	What Are the Clues? Patterns in More than One Direction	48
15	Wrist Bands	52
16	Take Two: Fair or Unfair?	55
17	What Shapes Can You Make?	58
18	Counting Cubes	63
19	The Orange Game	66
20	Darts, Anyone?	69
21	How Many Towers?	72
22	Talking Club	75
23	Fair Share	78
24	The Math of Motion	81
25	Marbles	83
26	What Can You Measure?	87
27	Which Graph Is Which?	91
28	Stair Skipping	94
29	Three-Way Sharing	97

ACKNOWLEDGMENT

This book is composed of a range of problems from the Problem Solvers column in *Teaching Children Mathematics*, NCTM's journal for elementary school teachers. The problems appeared over the course of three years, from 1997 to 2000. The solutions and children's work that are presented come from around the United States and—a few—from overseas. The authors would like to thank all the people who are using the problems posed in the Problem Solvers.

We would also like to thank Nordie Deal and Danny Breidenbach for their valuable input as the Problem Solvers developed and evolved.

INTRODUCTION

In the last few years, teachers of mathematics from prekindergarten through postsecondary school have become more aware of the central role that problem solving must play in the lives of all children. As such, problem solving needs to be more central to the mathematics curriculum for all children. That message has been supported by the recent National Assessment of Educational Progress (NAEP) data and the Third International Mathematics and Science Study (TIMSS) study. When we look at the NAEP data over time, they indicate that our students are increasing their mathematics skills. Although that news is encouraging, the TIMSS data are not so encouraging. In the area of problem solving and reasoning, the TIMSS data indicate that our students are not as strong as hoped. Children need many more experiences in problem solving to apply and strengthen their mathematical abilities in order to be lifelong users of mathematics. These abilities include the ability to reason, to communicate mathematically, to conjecture and test strategies that children are developing for solving problems, and to explore new and challenging problem situations without knowing exactly how they will solve the problem.

Since 1989, the National Council of Teachers of Mathematics has focused on five goals for all children. Problem solving is one of the five goals, and success in achieving that goal is intertwined with success in achieving the other four. State and national assessments reflect the change in curriculum that has placed problem solving in a more central role in mathematics. However, teachers need resources, support, and time to implement such dramatic change in the classroom. The purpose of this book is to offer teachers problem-solving situations that are intended to engage their students in interesting explorations in which they do challenging, interesting problem solving with significant mathematical content. It is also intended to support teachers by including examples of children's work, suggestions for how the teacher might pose problems, a range of questions and possible assessments for use with children, and by providing the backdrop of our own experience with some of the problems.

AN OVERVIEW OF PROBLEMS AND PROBLEM SOLVING

What Is Problem Solving?

"Problem solving means engaging in a task for which the solution method is not known in advance." *Principles and Standards for School Mathematics*, p.52

PROBLEM solving encompasses the acts of exploring, reasoning, strategizing, estimating, conjecturing, testing, explaining, and proving. It is a very active process for those involved. Through problem solving, we are challenged to think beyond the point where we were when we started. We are challenged to think differently. We are challenged to extend our thinking about a situation in a way that is new or different.

What Is a Problem?

A problem is a task that requires the learner to reason through a situation that will be challenging but not impossible. Within the problem there is a hurdle that the learner cannot immediately see how to get over or around. The learner must determine what strategies to use to come up with a solution. The strategy itself should not be immediately obvious to the learner, but it should be within the realm of strategies possible for her. An analogy is that of a mathematical journey that the learner is taking. The challenges along the way are the problem situations. The challenges require the learner to stretch herself in her thinking and in the use of mathematics. As the learner figures out ways to overcome the challenges, she develops strategies that prepare her for more difficult challenges encountered. Most often, the learner is working with others in a group to meet these challenges. Knowledge developed about how one met previous challenges is used to meet new challenges that are increasingly difficult. Along the way, the learner is developing confidence in the ability to meet the challenges and is becoming more experienced at successfully using strategies to meet new challenges. Teachers, the mentors along learners' mathematical journeys, are responsible for challenging learners in ways that allow them to grow as problem solvers.

There is a difference in the use of language here from what is traditionally known as "math problems." What many refer to as problems—a set of number sentences intended for practice in developing a skill—we refer to as exercises. As one practices the exercises, one becomes more "fit" at using the skill. What we refer to as problems are those that come from the body that many refer to as "word problems." In that body of problems, there are still some that are in the realm of exercise. For example, watch a child that has a sheet of five word problems that are all the same format with only the numbers changed and that can be solved by repeatedly using the same strategy. The first one or two may have been challenging problems. Once the child recognizes that the others can be done the same way and begins a routine of repetition, however, the problems become exercises that are being used to practice a skill. By the same token, once a child is experienced enough with problem solving, what was once a problem to that child may now become an exercise. Or, what is a problem for a learner who is newer to problem solving may be an exercise to a more experienced problem solver. Hence, a teacher who teaches with problem solving must be observant. When she sees problems become exercises, she must ask different probing questions such as "Does it always work?" or "What do you do if … ?" to keep a student working at a problem-solving level.

What Is a Good Problem?

A good problem has many characteristics, and there is no "formula" for determining what is a good problem. Some of the common characteristics of good problems are shared here. A good problem must be challenging to the learner. It should hold the learner's interest. It should be a problem that the learner can connect to her life and to a range of concepts in mathematics or to other disciplines. A good problem for a class should also have a range of challenges embedded in the problem for learners who are at different levels mathematically and as problem solvers. A good problem should also have several ways that it can be solved.

From reading the list above one might be amazed that we could share even twenty-nine good problems. But there are many good problems available in a range of resources for teachers to use in their classes.

Also, the notion of whether a problem is good is relative. Because several of the characteristics above depend on the learner, what is viewed as a good problem for one learner may not be a good problem for another. This fact makes it difficult to list problems by grade level. Whether one's class is multiage or composed of children of the same age, the problem-solving experiences of the children will be varied. As a result, it takes well-chosen problems to challenge all the children without frustrating some with an impossible task. We have therefore provided a matrix at the end of this section with a range of grades in which a problem could be used. When selecting problems for the Problem Solvers, it was important to choose those that would be good problems for many children. We looked for problems that would be accessible to many and would still be challenging to many.

What Does It Mean to Be Successful at Problem Solving?

Having success means that the child has experienced a way of thinking about mathematics that they had not known before they encountered the problem. Thus, whether a child is successful depends a great deal on their level of problem solving as they encounter the problem. Success will involve the process of problem solving as well as understanding the content presented. In addition, success might mean more facility with one or more of the different strategies used in solving problems. For example, when solving problem #3, How Many Rectangles (HMR), there are many places where children can have success, whether or not they arrive at a final answer. If a group is working on HMR and they recognize that they must resolve the question about whether to include the squares in their count of rectangles, they have had success with the problem. The relationship between squares and rectangles is a very important one for children to explore. When they begin to understand that squares meet all the criteria needed to categorize them as rectangles, they have made a strong connection. This is one of the ways children can be successful with HMR.

Another way that children can achieve success with HMR is to begin to see the rectangles embedded in the diagram, so they see that there are several sizes of rectangles present. Even if children find some of, but not all, the fifteen different sizes, they have had success. They have begun to look at relationships between parts of figures in a different way, thus expanding their spatial awareness.

A further level of success can be achieved when children devise a way of counting the rectangles. There are many ways that the rectangles can be counted and represented, both symbolically and using models. As children develop ways of counting, they are successfully expanding their number sense. This will serve them well with future problems, both in school and in applying mathematics in their lives.

Yet another way that children can achieve a level of success with HMR is to find patterns in the number of different-sized rectangles that they are counting. For example, a child who (a) conjectures that as one moves from counting 1×4 rectangles to counting 1×5 rectangles, there will be one less rectangle in the 1×5 category than in the 1×4 category, and who (b) tests that conjecture, has been successful at pattern recognition and theorizing about patterns, thus further developing his level of number sense.

There are even more layers of success possible when exploring HMR, and the solution has not yet been discussed. If, in each of the instances above, a child thinks about the concepts embedded in HMR differently because of the experience with the problem, then that child has had success with the problem.

One of the important benefits of focusing on all the areas of success rather than only a right answer is that we change our focus. We focus on the learning rather than a single answer. The learning takes place while the experiences above are occurring, rather than when an answer is written down. Thus, a student who has success with several of the aspects of the problem, but does not arrive at the answer still has learned a great deal as a result of exploring HMR. It is the process that holds the opportunity for learning rather than the answer.

The Teacher's Role in Problem Solving

The teacher's role in problem solving requires that several hats be worn—something teachers do well. First, the teacher plays a crucial role when choosing meaningful problem-solving tasks. When selecting a problem, the teacher must analyze and sometimes adapt a problem, anticipating the mathematical ideas that can be brought out by working on the problem, and anticipating students' questions. Teachers can decide if particular problems will help further their mathematical goals for the class (NCTM 2000, p. 53).

Then the teacher must determine how to present the problem for students to solve. It must be presented in a way that will interest and engage the students. It must be presented so that all students believe that they are challenged but that none think the task is impossible. In this phase, the teacher decides whether to have students work collaboratively or individually. The benefits of having students collaborate are many and the drawbacks are few. Individually, many students give up very quickly when they first begin doing problem solving. If they give up, there is no way to engage them in order for them to become better problem solvers. In groups, they have much greater confidence that they can come up with strategies. They tend to work on a task longer, which increases the likelihood that they will be successful. Another benefit of working in groups is that students hear a broader range of strategies than they might come up with individually. They then adopt many of these strategies to test and use in problem solving, thus extending their own understanding. Students also enjoy working together to solve problems. This is another reason they will remain involved in the task longer. They will also be more likely to remember more of what they discuss with someone else than they will if they just jot down some ideas and move on to the next problem.

One of the points that is often raised as a drawback is the concern that some students are not as engaged or "not working as hard" as others. In most instances in life some people are more involved than others. However, the authors have found that most students tend to be more engaged and to work harder in a group than when they are on their own. For those few that are less productive in a group, room can also be made for them to work alone. Another point that is raised as a drawback is that the noise level in the classroom is heightened because students are talking about the mathematics they are learning. That mathematical communication is one of the five goals of mathematics teachers. The conversations are a rich and meaningful part of the learning process. A good way to determine a reasonable level is to ask students to speak so those in their group can hear them and not much louder.

If the students work collaboratively, it is a good idea to give each student about 3–5 minutes to work on the problem and try a few strategies before the group comes together to work on the problem. This avoids the problem of having one student see a strategy immediately and tell the others how to solve the problem before the others even have a chance to think of what the problem is asking.

After the decision is made about how to present the problem, then the teacher has the hardest task of all. She must allow students to wrestle with the problem, without just telling them the answers. We teachers are a very helpful group of people. But in the process of helping, we can go too far and actually do the thinking for the student. If we reach the point where we're just telling them what to do, they are not engaged in the process. At this point the situation is no longer a problem situation for the students. The teacher's role at this point is very challenging. Knowing one's students and knowing when to step in is crucial. We don't want students to reach the point of no return with frustration. But by the same token, we want them to do the thinking, not the teacher. That's one reason why it's helpful for students to work in groups. If one idea is attempted and failed, there is usually another one at hand. No one knows, however, if the new idea will work, so everyone must use critical thinking to test the new idea.

The final role of the teacher is to determine how to assess what the students are learning as they are doing problem solving. This allows the teacher to determine how well students understood the task, and it guides teacher's decision making with regard to further problem solving. There are several ways to assess students' understanding of problem solving. A few general ones are given here.

One way of assessing understanding is to listen to the conversations as the students are solving the problems. Recording ideas that indicate a student's understanding is a good way to keep track of what is being learned. A second way is to have them explain—in a letter to an absent student or another friend—what the problem they solved was about and what strategies they used successfully and unsuccessfully to solve the problem. They can also explain how they know if their answer is reasonable or makes sense in the situation. A third way is to give them an extension that requires them to use what they learned in the original problem in order to solve the extension. This third form of assessment provides a paper trail but there are some drawbacks to grading it. Developing understanding takes time. Applying what was just learned in a new situation is challenging. Learning may not immediately be transferred to the new situation, even if the student understood it in the first situation. Often, it takes time and many experiences for the new understanding to become part of usable knowledge. Thus, the assessment may be more about how someone responds in a new situation rather than assessing what was understood about the original problem. Extensions are excellent ways to develop further understanding of the problem situation, but using them as tools for assessing the original problem can be challenging.

The many decisions the teacher makes about how and when to include problem solving in mathematics are important to helping students be good problem solvers. The more regularly that teachers make it part of the curriculum, the more opportunities students will have to become successful problem solvers.

Learning Mathematics through Problem Solving

It is important for teachers to understand why problem solving should play a role in their classroom in order for them to devote the time and energy of their class to problem solving. As students, many of us learned twelve years and more of mathematics that was just memorized bits and pieces. Mathematics was disconnected from our lives, and we had no clue about how to apply it. Indeed, we may not have even known that it could be applied.

As teachers, we know that all mathematics can be applied to people's lives today and into the future. When children learn the mathematics curriculum through problem-solving situations, they learn how to apply the mathematics as they are learning it. They can make connections within mathematics and to other areas of the curriculum, as well as to the world around them. That is a powerful reason to learn mathematics through problem solving.

Another reason to learn mathematics through problem solving is so that students can understand what they have learned. They are also more able to remember what they have learned if they were actively involved in learning it. Problem solving is an active way to learn mathematics. Students are interested in being involved and want to explore ideas further. They are able to reason and connect ideas in ways that continually amaze those around them.

As you read through the problems and examples of how children of many ages solved the problems, you will see the power of learning mathematics through problem solving, and you will see how children are problem solvers.

References

National Council of Teachers of Mathematics (NCTM). *Principles and Standards for School Mathematics*. Reston, Va.: NCTM, 2000.

THE following table was designed to aid teachers in selecting problems to use with their classes. Although the information is intended to help you make better choices about what problems to use, the grade levels suggested are only estimates. As mentioned above, the teacher must decide whether a problem appropriately challenges the children in his class. Reading the description of how other children solved a problem will certainly help you decide whether a problem is a good fit for your class. It is recommended that the teacher err on the side of challenging the children more rather than less. Because we don't want to lose our children, we don't always challenge them enough. We encourage you to push the envelope to the level that you think your students will be doing some good problem solving.

Problems by recommended grade level and content focus

	K–2	3–5	6–8	Number	Patterns/ Algebra	Geometry	Measurement	Data & Probability	Reasoning & Proof	Communication	Connections	Representation
#1	•	•		•	•	•	•		•	•		•
2	•	•	•	•	•				•	•	•	•
3	•	•	•	•	•		•		•	•	•	•
4	•	•	•	•	•			•		•	•	•
5	•	•	•	•	•				•	•		•
6	•	•		•	•				•	•	•	•
7	•	•		•	•				•	•		•
8	•	•		•	•		•		•	•	•	•
9		•	•	•	•				•	•		•
10	•	•		•	•				•	•		•
11	•	•	•	•	•	•			•	•		•
12		•	•	•	•				•	•		•
13		•	•	•	•		•		•	•		•
14	•	•	•	•	•				•	•		•
15	•	•	•	•	•			•	•	•		•
16	•	•	•	•	•			•	•	•		•
17	•	•	•	•	•	•			•	•		•
18	•	•	•	•	•				•	•		•
19	•	•	•	•	•		•	•	•	•		•
20	•	•	•	•	•				•	•		•
21	•	•	•	•	•				•	•		•
22	•	•	•	•	•			•	•	•		•
23	•	•	•	•	•		•	•	•	•		•
24	•	•	•	•	•	•			•	•		•
25		•							•	•		•
26	•	•	•	•	•	•	•	•	•	•		•
27	•	•	•						•	•		•
28	•	•	•	•					•	•		•
29	•	•	•		•				•	•		•

Note: An "•" in a box means that some aspect of the problem pertains to that category.

Problem # **Title**

1 .... Animals and Fences at a Zoo
2 .... Code It Be?
3 .... How Many Rectangles?
4 .... Double That Dough
5 .... How Much Film?
6 .... When Will We Reach 1/2?
7 .... Decoration Delight
8 .... You Gotta Have Heart and Blood!
9 .... Sandwiches
10 .... Shuffling a Line
11 .... Counting Squares
12 .... How Many Times Can You Take 1/2?
13 .... Guess the Weight
14 .... What Are the Clues? Patterns in More than One Direction

Problem # **Title**

15 .... Wrist Bands
16 .... Take Two: Fair or Unfair?
17 .... What Shapes Can You Make?
18 .... Counting Cubes
19 .... The Orange Game
20 .... Darts, Anyone?
21 .... How Many Towers?
22 .... Talking Club
23 .... Fair Share
24 .... The Math of Motion
25 .... Marbles
26 .... What Can You Measure?
27 .... Which Graph Is Which?
28 .... Stair Skipping
29 .... Three-Way Sharing

1
ANIMALS AND FENCES AT THE ZOO

The Problem

A zookeeper was promised that she could have some special animals called *mathemals*. She has twenty connecting cubes to be used as fencing to build a pen for the mathemals. What type of pen can she make to hold the most mathemals?

She must follow these rules:

- She must use all twenty connecting cubes to build the pen, with each cube joining another cube, face against face.
- The pen must be closed, with no doors or openings, so that the mathemals cannot get out.
- Mathemals cannot be allowed to stand on top of one another in the pen.
- Each mathemal in the pen uses the space of one cube.

Some teachers may wish to make the problem more open-ended by dropping the face-against-face constraint, allowing students to determine for themselves how to meet the requirement that the pens have no gaps or openings.

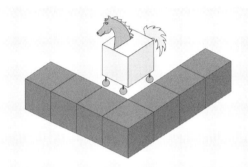

The materials needed for the project are these:

- Twenty connecting cubes of one color to use as fencing and a large supply of connecting cubes of another color to use as mathemals when testing various solutions. Connecting cubes are suggested, as they may be helpful for younger children.
- Paper for recording the results.
- Grid paper in the same dimension as the cubes being used.

Range of Possible Teacher Questions

- Are there any gaps in your pen?
- Do you see a relationship between the arrangement and the number of mathemals?
- Can you use any ideas from those around you to fit more mathemals in your pen?

Where's the Math?

The math in this problem situation is very rich. Students are generating patterns as they look at different configurations for the pen. They are also using spatial sense as they stretch their thinking to find the largest pen. They are using measurement as they determine what is meant by "largest"—the largest area covered by a perimeter that is fixed. Students are using problem solving throughout the activity. As they discuss their interpretations of the problem, their thinking as they work, and their solutions, students are communicating mathematically.

SAMPLES OF HOW THE PROBLEM WAS SOLVED BY OTHER CHILDREN

SEVERAL teachers presented the "Animals in a Zoo" problem to their classes. Rena Mincks from Pullman, Washington, used this problem in her first-grade class and reported the following:

> I introduced the problem just as written. At first the room was very silent as students became engaged in the activity. As I walked around the room I heard such comments as "This is hard," "I didn't use all 20," and "I think I see a pattern here!" My students were using many of their already-learned skills and strategies.
>
> The criteria I used to assess students were as follows: Could they count out twenty blocks? Could they follow the rules? Did they accept their first solution, or did they continue to search?

Mincks gave her students grid paper on which to work with Unifix cubes. The students then transferred their answers to the sheets shown in figures 1.1, 1.2, and 1.3.

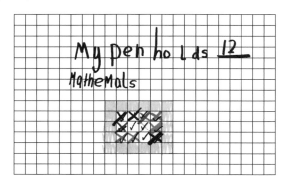

Figure 1.1

Figure 1.1's response was given by four students, although the fence was oriented differently in a few cases. These students had the correct idea of a fence and used twenty cubes to build it, but they did not build the largest pen. Four students used a five-by-seven rectangular fence to hold fifteen mathemals.

Some students had a solution of twenty. Their reasons, however, were different. One student, whose answer is shown in figure 1.2, tried to extend the constraint of fence to find the largest area.

Two other students' rectangular fence held twenty mathemals. Their conception of the problem focused on getting the pen to be as large as possible and, in so doing, used twenty-two cubes for the boundary. One student combined these two strategies, ending up with twenty-four mathemals inside a nonrectangular fence with twenty-three cubes.

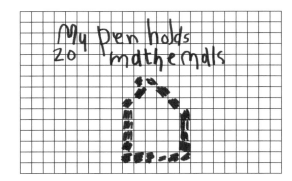

Figure 1.2

Mincks observed, "Overall I noticed that many children found a solution and quit there. They had an answer, and that was all they needed. A few students tried to be creative but soon realized that it was difficult to do and stay within the rules as written in the problem. Several of the students kept trying and trying different solutions until they finally settled on one 'best' solution."

Four students built the largest pen possible under the constraints given. Figure 1.3's example is especially important. Note that the student made three errors in writing the numerals yet clearly was able to complete this problem-solving effort successfully. Although still developing writing skills, this learner was quite capable of solving a problem that required analytic thinking. As a result, the child met all the criteria for success as established by the teacher. Only one child was not able to write something meaningful about this problem. Mincks concluded, "This task was enjoyed by all my students. It invoked problem solving and reasoning. It allowed students time to practice their communication skills and encouraged identification of connections among

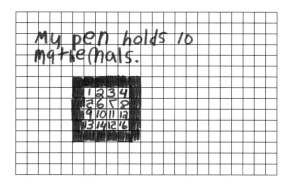

Figure 1.3

various mathematical concepts and principles. This was a very insightful activity. I learned a lot about my students."

Clearly, these first graders were successful in this problem-solving effort. Almost all understood the problem and were able to make substantial progress toward a solution. The next examples show that children used different constraints on their fences and arrived at different conclusions.

Stacy Baker of Peoria, Illinois, who tried this problem with her third-grade class, reported, "The students interpreted the first rule to mean that the blocks were touching, not back to back. Since the whole class thought this way, I let the rule stand. On the first day, I let them experiment and record all their ideas. At the beginning of the second day, I had them each post on graph paper their biggest area and let them work from there. As a follow-up, I let them create their own mathemal and write about a day at the zoo."

Seven students used the approach shown in figure 1.4. The shape of their design moved from a rectangle to a shape approaching a parallelogram—which the students called a diamond—with various interesting shapes ranging from a low of sixteen to a high of forty animals in the pen. One student, reflecting on the two-day process, commented, "What I did different was I made a diamond and got thirty spaces all together from yesterday." Another student's expression is found in figure 1.5.

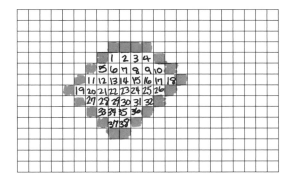

Figure 1.4

Although the rules had been changed and agreed on, four students still tried to make a rectangular pen in keeping with the original problem, as shown in figure 1.6. One of the four students ignored the twenty-block-boundary constraint.

Two students used modified directions to make the largest pen possible, one holding forty-one mathemals; and one other student's pen contained forty mathemals. These three students showed remarkable insight in the two-day process that resulted in their solutions. Figure 1.7 shows just one example.

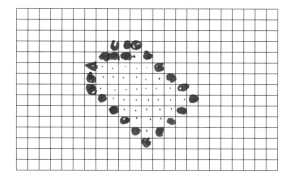

Figure 1.5

We see that these students were very capable problem solvers who understood the task and were able to develop and refine their product successfully over a two-day period and to explain their thinking through writing.

Figure 1.6

Barbara Kelly of Galesburg, Illinois, from whom we obtained the problem, also had her first-grade students complete this task. She worked with her students to ensure that they each knew that they could work with twenty cubes. She also gave each child eighteen multilink cubes of a different color to represent the animals, in this case, cows. She gave specific instructions for the pen by saying, "Your pen has to be closed, no doors, so the cows don't get out." Here are some of her observations and comments:

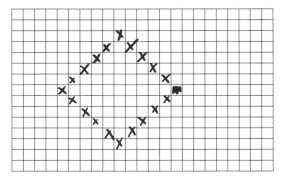

Figure 1.7

I walked around and listened while they worked. I heard "I need more cubes," from children who had constructed a pen that was not closed and "I have extra cubes" from the child who constructed a pen with less than the twenty provided. I told them that everyone has twenty cubes and that's all they can use.

It took some children just a few minutes to construct a pen that fit the qualifiers. One child asked if he could build two pens. I told him that if he thought that they would hold the most cows, then that was okay with me. He abandoned that idea after working on it awhile and determined that the one pen would hold more. Others became frustrated and looked at neighbors to confirm that a solution was possible and get ideas for completing the task.

As children finished building their pens, I gave them paper and asked them to draw a picture of their pen and write about it. When the recording was done, I asked them to bring their drawings and pens to the floor for sharing. We recorded the results on chart paper.

The children's results are shown in this table.

Number of Cows in Pen	Number of Children with That Count
6	1
7	1
10	1
12	4
15	4
16	4

We looked at the pens that held 16 [the most] and compared them to the pen that held 7. I asked them what they noticed about these pens. The group made two observations:

Skinny pens hold fewer cows.

Square pens hold the most cows.

Then I asked the children to try to use their cubes to make the pen like the one that held the most. I thought this would be an easy task, since several models were available to look at in the circle. But as I watched, I saw that this was not an easy task for most of them.

What Are the Students Telling Us?

The experiences of these three teachers provide some good insights into the work that children can do, for example:

- Children are able to handle nontraditional problems.
- Having children negotiate rules or asking for input from students can help determine the structure for the problem, although doing so may change the format of the problem and produce a different result than expected.
- The necessity of communication in mathematics is highlighted. Students were asked both to model and write their answers. The students' writing sheds insights into their thought processes that the pictures do not.
- Assessments may evolve with the children's understanding. For example, allowing students to work on a problem for several days allows for deeper exploration. An assessment of most students in Stacy Baker's class on the second day would have yielded quite different results from one on the first day. A delay gave children more time to refine their thoughts.
- Primary children, even in grade 1, can grapple with and solve problems involving sophisticated mathematical content. This problem as posed is one of the classics involving changing area while maintaining a constant perimeter. This exploration need not be reserved for older students.

2
CODE IT BE?

The Problem

A teacher was cleaning for a rummage sale. In one container he found over 500 each of letters A, B, and C. He decided to use these letters to create a three-letter code name for each person in his class, adults as well as students. Are there enough different combinations so that all of the students and adults will have their own three-letter code name?

The materials needed for the project:
- Numerous cards containing letters A, B, and C
- Paper for recording results

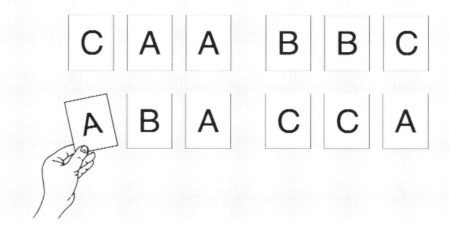

Range of Possible Teacher Questions
- What is meant by "different combinations?"
- Do you see any patterns?
- How can you tell if you have all the possible combinations?
- Explain how you figured out a strategy for solving the problem.

Where's the Math?

In solving this problem, students will be engaged in several mathematical activities. They will be looking for patterns. They will be determining how many combinations can be generated from a finite number of items. They will be involved in problem solving. And they will be communicating mathematically.

SAMPLES OF HOW THE PROBLEM WAS SOLVED

STACY Baker from Pleasant Hill #69 in Peoria, Illinois, tried the problem with her third grade class as it was posed in the May 1997 issue of *Teaching Children Mathematics*. Students began a rich discussion of what the problem meant and what they needed to determine prior to trying to solve the problem.

"Immediately questions arose such as how many students and teachers were in the (fictitious) class. My class decided to model it after our own class. We have 27 students and 3 teachers at various times during the day."

The class also engaged in conversation about what codes would be considered "the same" and what codes would be classified as "different." "They wanted to know if they could work backwards. This means is ABC the same as CBA? After class discussion they decided that they were not the same and could both be used. They also wondered about AAA. The class again decided that this was acceptable. Later they wondered if it had to have three letters. Some tried to use small and large combinations, but they decided that the directions specified three letters that were all capitals." This conversation about a problem and about problem solving in general brings much more meaning to a problem solving situation for the learner.

Once the class decided on some ground rules, the students generated their lists with a variety of strategies. Some students attempted a list by exhaustion with no apparent trends or patterning. Often those students were not successful in naming all of the different codes, although their lists had 30 code names on them. For example, one student had 34 code names but at least nine were repeats.

Other students used trends and patterns. Some students were very sophisticated in their approach. Some common approaches have been noticed and named. Examples of students' work are included to illustrate an approach. One strategy which was commonly used to some degree was that of listing triples, doubles and singles in groups. For example, one student's approach was the following:

AAA BBB CCC
(triples)

ABA ACA BAB BCB CBC CAC AAB AAC BAA
CAA BBA BBC CBB ABB CCA CCB BCC ACC
(doubles, where a letter repeats)

CAB CBA BAC BCA ABC ACB
(singles, no repetition of a letter)

Her codes were recorded in a single column, in the order given. They have been separated in order to highlight her grouping process. However, they remain in the order in which she listed them. Also, strategy names have been added in parentheses. Within her strategy of listing all the triples, then all the doubles, then all the singles, she used two other strategies common to many students. She used trends within her doubles to create a more thorough list. First, she listed all the doubles with the different letter in the middle. Then she listed all of the doubles that had the different letter on one end or the other. The other strategy she used was pairing, where one letter was held constant and the other two were interchanged, from one code name to the next. This very sophisticated combination of strategies on the part of this third grade student and others helped them succeed in writing all 27 of the possible code names. Other students who listed all 27 code names without including repeats incorporated some combination of the above strategies. Several students would change strategies as they went through the process of listing. For many students, the changing of strategies didn't seem to keep them from listing all of the names.

Another strategy used by a few students was to list all possible code names that began with A, then all possible code names that began with B, and finally all of the ones that began with C. They met with varying degrees of success and seemed to be able to organize thoughts fairly well using this approach.

One interesting aspect of the situation in Stacy Baker's class was that, since the students decided on 30 as the number of code names they wanted to generate, some students felt obliged to produce 30. They listed all 27 different names, then added three that they thought they missed, although they were duplicates. They felt the need to find 30, although there were not 30 different ones.

Ms. Baker's overall assessment of the process was, "I was amazed at how quickly they worked on this problem and how organized they seemed to be. They really enjoyed it." Her approach to grading the student work was to look for three levels of understanding.

Plus:	Correct answer and explanation
Check:	Some work and no explanation
Minus:	Wrong answer and no explanation

What makes the above experience interesting is the added issue that the number of different codes needed is not possible with three letters. Some students were comfortable with stopping at 27. Others felt the need to "complete the given task" and as a result, they repeated codes. This sort of experience is important for learners to grapple with in order to recognize that the task is not always possible. Not everyone in the group of 30 can have their own code if there are only three letters to choose from and each code has three letters in it.

Sue Holzwarth's 3rd grade class at Lincoln Elementary in Macomb, Illinois, also worked on a version of the problem. She posed the problem this way.

> I was getting ready for a garage sale when I came across a box with over 500 each of the letters A, B and C. If I were to create a three-letter code for each person in our class (24) and for myself, would there be enough letters?

Ms. Holzwarth allowed students to model the problem as an option. "We had just done trains of three using two colors of the unifix cubes. I told the students they could use the cubes if they wanted, but they had to write the letters in the codes so we could check them. Most students who were with partners associated very quickly with the previous problem." This observation was viewed as being very important to the process. "Students set to work using the cubes and trying to see how many different combinations they could get. It was different from their other assignment because there were three choices."

One of the pairs of students did a sort of one-step change process. They changed by one attribute with each successive code they wrote. Upon exhausting all possible combinations of a type with the one-step process, they would go off in a different direction. The pair began this process after writing down the first few codes. They had duplication six times and didn't have all possible combinations listed. This could have been due to the direction changes that took place. They generated their patterns using the blocks and attached the letters after they created all of their three-cube combinations. They had 28 codes with repetition.

ABA ACA CAA AAA BAB CCC BCB CCB CCA
BCA BCC BAC BBA ABC CAC ABB BCB ACB
CBB CBA AAB CBA CAA CCA BCA ACC BCA
ACB
(One-step change process mentioned above.)

Another pair recorded their codes using a different color for A, B, and C. They generated 25 of the 27 different codes. They generated enough names for all of the students in the class to have a different one. Their overall strategy was a version of listing triples, doubles and singles, and going back to pick up what they missed. They listed pairs within the larger strategy. They wrote their responses in column form on grid paper. The As were all purple. The Bs were all blue and the Cs were all yellow. The color-coding made it easier to scan for duplicates.

A	B	C	A	B	B	C	A	B	B	B
A	B	C	B	A	C	B	C	A	C	B
A	B	C	A	B	B	C	A	A	C	C
C	B	C	B	C	A	A	C	A	B	C
C	A	A	B	A	C	A	B	A	C	B
B	C	C	A	B	B	C	A	B	A	B
C	A	A								
A	B	B								
A	B	C								

One of the children described their approach in the following way. "We guessed and checked. We also did opisons (opposites), and put the opisons in pairs." It is interesting that they didn't focus on describing the trends but rather on the opposites. This indicates that they may not have been aware of how they were generating their list. Or, it may indicate that the trends of triple, double and single may have been a coincidence. This pair also kept score of who generated more codes and circled the number next to each name.

One pair of students used the color coding on grid paper after the first twelve codes were written. They had four duplicates, but none of them repeated in the color-coded section. They were all repeats from written form to color-coded form, where the visual was not present to begin with.

A group of three students had the following list.

1. AAA   12. CAB
2. ABC   13. CBA
3. ACB   14. CCA
4. AAB   15. CCB
5. AAC   16. ACC
6. BBB   17. BAA
7. BAC   18. BCC
8. BCA   19. CAA
9. BBA   20. ABB
10. BBC   21. CBB
11. CCC

Upon examining the list as it was given above, it appeared that the students had done only the A then B then C strategy of listing all possible codes beginning with each letter. Their explanation revealed a more involved process. "We tired each one. Then we turned everything around." If the list is reorganized as follows, one can see the process of turning things around which they referred to. The items are numbered to follow the order in which the students wrote them.

1. AAA	6. BBB	11. CCC
2. ABC	7. BAC	12. CAB
3. ACB	8. BCA	13. CBA
4. AAB	9. BBA	14. CCA
5. AAC	10. BBC	15. CCB
16. ACC	17. BAA	19. CAA
20. ABB	18. BCC	21. CBB

The first five indicate where they tried each one. Six through ten and 11 through 15 are results of turning everything around. Then they began the process of trying everything with 16 and turned each code around before going on. Had the explanation not been there, the pattern might not have been detected.

There were pairs who attempted the list-by-exhaustion method. They met with the same results as the students in Sue Baker's class. They often had several duplicates and wrote comments about not having enough time to complete the task.

Sue Holzwarth asked students to write their answer and the way that they found it. She commented that her students "resist with a passion when it comes to writing." She said that most students explained that they used guess and check, although she felt that building the models and transferring them to paper was much more common.

Lisa Shaffer of East Meadow, New York, tried the problem with her fourth grade class. She posed the problem in a similar way to Sue Holzwarth. When asked "Are there enough combinations so that all the students will have their own three-letter code name?", some students immediately responded, "Yes." Ms. Shaffer asked the students why they thought this. "They said they knew it would be enough because the teacher had over 500 letters. They thought since 500 is a large number there would surely be plenty."

One student recognized the need for further discussion by pointing out that there would need to be more information provided. "How are we supposed to know if we have enough letter combinations if we don't know how many kids are in the class?" The group then decided to revise the problem so it would read, "Are there enough combinations so that all 34 students will have their own three-letter code name?"

As Ms. Shaffer walked around the room, listening to discussions, some of the comments she heard were:

"We have to find a pattern."

"I wonder if we can multiply to find the answer."

"I'm starting with all the a's."

"This can be confusing. I keep thinking I have a new order but then I find out I wrote it already."

"I think there are 27 combinations because I started with all the A combinations, then I did all the B combinations and C combinations. There are no more to do and I only have 27."

Her students used the strategies of listing all of the As then all of the Bs then all of the Cs. Some of the students listed triples then doubles then singles. Some students came up with more than 27 by having repeats. Again, using a strategy and organizing seemed to aid in avoiding duplication of a code.

The teacher who originally shared this problem with the editors, Barbara O'Donnell of Knoxville, Illinois, gave as one of her objectives that "Students will devise a systematic way to list combinations of events, items, or objects." and that "Students will be able to identify the strategy, Make an Organized List, as the tool used for this problem."

Her analysis of the results with her fourth grade class were that :

- Six students found all 27 combinations with an organized list.
- One found all 27 combinations without an organized list.
- Six found some of the 27 combinations with an organized list.
- Two found some of the 27 combinations without an organized list.
- One found some of the 27 items but used some items more than once.
- And one attempted a list, but confused the categories.

Ms. O'Donnell made an interesting point about the results in her fourth grade class. "Many of the students thought there were 9 ways. This could be due to the fact that we had discussed a short cut. That is that if you can find all of the combinations for one of the first group members, you could just multiply by 3 to include the combinations for the other two." The assumption that the short-cut must work may have kept students from exploring further.

Her reaction to an assessment of the objectives was as follows. "All in all, I am still happy with the results. At the onset of teaching this strategy, the students seemed to have no prior knowledge of the concept of Making Organized Lists to solve problems. When giving this problem to the class, I did not mention the strategy. They assessed the need to use the strategy and employed it."

This problem was attempted with third graders and fourth graders in groups of two to four students. The results indicate that most students met with a degree of success in understanding and solving the problem. They also indicate that some of the students employed a systematic approach to organizing the codes, to ensure that the desired results would be achieved.

What Are the Students Telling Us?

- Students in the primary grades are capable of rather sophisticated organizational skills in approaching tasks.
- There is a need for students to experience problems for which the results cannot be predicted. There is a need for the problems posed in the classroom not to always have an answer so that when the question "Is it possible?" is asked, the assumption is not that the answer is "yes." Some students were confident enough to say that there were a certain number of combinations and no more. Others assumed that there must be the number the class had chosen and lost sight of the constraints because of that.
- Students were able to jump right in when they and a partner worked together to discuss the problem.
- Discussion among all the students seems to lend itself to greater "ownership" of the problem.
- There is always a need for mathematical communication. The group that explained their approach by saying, "We tried each one. Then we turned everything around." indicated a strategy that was not obvious unless their comments were included.

As a final note, it would be interesting to pose the following problems to students as another look at this task.

Given three letters for use, how many codes can be generated?

If we have the three letters X, Y, and Z, how can we organize the code names so we're sure that we've listed them all?

Thanks to the teachers who submitted their results and their comments for us to share with the reader. Please try any of the problems with your classes and share the results with us.

3
HOW MANY RECTANGLES?

The Problem

How many rectangles appear in the figure below?

Range of Possible Teacher Questions

- Is a square a rectangle?
- Are all the rectangles the same size?
- Have you found any hidden rectangles?
- Can rectangles be overlapping?
- Do you see a pattern in how many of the different sized rectangles there are?
- How can we test to see if the pattern holds?
- Is there a way you can use a pattern to sum the different-sized rectangles?

Although the students may need to discuss what the problem means, and what is acceptable to count as rectangles, we encourage you to avoid giving too much guidance. It is important to view this task as more than an exercise for which students are seeking a correct answer.

Where's the Math?

In this problem, students will explore figures embedded within figures. They will explore properties of squares and rectangles. They will have an opportunity to have a good discussion about whether a square is a rectangle. As students explore the model and verbally make conjectures, they will use and develop their spatial reasoning skills and strengthen their ability to communicate mathematically. Generating the number of rectangles of a given size will allow children to explore and expand their number sense. Finding a pattern for summing the different-sized rectangles will provide opportunities to explore number theory. If they generate a rule for summing the different-sized rectangles they will be using algebra. And extending the number of squares used will allow the children to conjecture about what would happen if the pattern was extended.

SAMPLES OF HOW THE PROBLEM WAS SOLVED

Responses to this problem revealed variety in how the problem was presented, differences in constraints imposed by students or teachers, and a wide range of successful strategies used in solving the problem.

Jennifer Temple, a preservice elementary teacher working with seventh graders at Audubon Elementary School in Rock Island, Illinois, modeled the problem using five cubes of different colors. The students were able to refer to the position and color interchangeably. After one student discovered how to count the rectangles embedded in the original figure, he shared a couple of examples with the others, who then generated their own. "It was as though the light went on and they could see all the other rectangles," Temple commented.

Many students needed to grapple with whether the squares making up the original figure should be counted as rectangles. Although many students counted the squares as rectangles, another large group counted only nonsquare rectangles, yielding an answer of 105. Still other students gave two answers, one including squares in the total, the other not including squares. Amy Robertson of James River Elementary School in Williamsburg, Virginia, stated that the issue of whether to include squares in the count was the first thing her class of fourth and fifth graders discussed after reading the problem.

Rhonda Kinnish of Chestnut Hill Elementary School in Midland, Michigan, dealt with this issue from the beginning. She began by placing attribute blocks on the overhead projector and having the class name the shapes. Before she presented the problem, they discussed the attributes of a rectangle and a square. Students generated definitions, and then they compared them with their textbooks and with a dictionary. Their definitions did not match those in the books, so the class discussed the definitions until they agreed that a square is a rectangle with the added attribute of having four equal sides.

In instances in which students did not discuss what the problem was asking, different interpretations of the question resulted. Many students from John Mirabelli's class at Sheppard Public School in North York, Ontario, counted how many *different* rectangles they saw. They established the notion of *different* as having a different length. These students saw 15 different rectangles and drew each of the different-sized rectangles. It would be interesting to repose the question to students who answered in this manner. Since they recognized all the different sizes, they had resolved one of the issues in solving the problem and should then be ready to move on to the next.

Some students appeared to be operating under the constraint that no two rectangles could overlap and found a strategy for solving the problem with this point in mind. In Seoul, South Korea, Minho Kim and Mangoo Park posed the problem to a class of third-grade and a class of fifth-grade students. These grade levels are equivalent to fourth and sixth grade in the United States. One student operated under this no-overlap constraint and produced the solution shown in figure 3.1. To find how many rectangles of a given length could fit in the original figure, she uses division while ignoring remainders. Note her grouping strategy in finding the final sum.

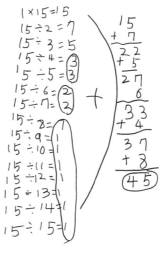

Figure 3.1. Finding the number of nonoverlapping rectangles

Many approaches were used by students who interpreted and solved the problem without any constraints. Figure 3.2 shows an example from Sara Jenkins's fifth-grade ESL class at Lattie Coor School in Avondale, Arizona. Emily uses colored brackets on grid paper to represent the rectangles in the figure. Emily stated that as of counting the 1×10 rectangles, she saw a pattern and did not need to continue her diagram.

Others used a sliding diagram to make sure that they accounted for all the rectangles. V. Snyder of Montgomery, Alabama, observed how one of her

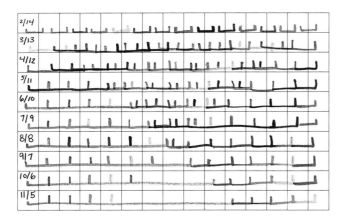

Figure 3.2. Emily's colored brackets represent the rectangles she saw.

sixth-grade students solved the problem. "He started with the first square and counted it as one [rectangle], looked at the first two squares together as the second rectangle and counted it as two, then included the first three squares for three, and continued until he had counted fifteen for his first count. Then he moved to the second square and continued counting using the same method that he had used for the first trip through. This time he counted fourteen as he moved through the rest of the rectangles." The sum of 15 + 14 + 13 + 12 + ... + 4 + 3 + 2 + 1 represented different ideas, depending on the approach used. The 15 in this instance represented fifteen rectangles formed beginning with the leftmost square. In another approach, 15 represented the number of 1×1 squares.

Marisol, from Jenkins's class at Lattie Coor School, did not attempt to calculate a specific number of rectangles, but she shows a general sense of the meaning of the problem. In figure 3.3 she uses a version of the sliding-diagram approach to show how many 1×10 rectangles exist in the figure, but she takes her pattern to an erroneous extreme. It would be interesting to have her explain her thinking.

Kristen, from Kinnish's class in Chestnut Hill Elementary School, found all 120 rectangles by combining the sliding strategy with an organized labeling system for describing rectangles. After numbering the squares in the original figure, she referred to the rectangles by the position of the squares on each end. She labeled her two 1×14 rectangles as

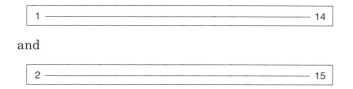

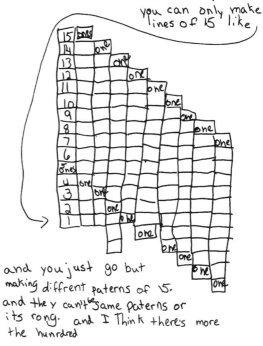

Figure 3.3. Marisol's attempt at the problem

(see fig. 3.4). She continued this process until she recognized the pattern while counting the 1×10 rectangles. At that point, she said, "It keeps going up and up following the same pattern through 15."

Ryan approached the problem simply by making an organized list. He wrote his pairs of beginning and ending numbers in the form 1/1, 1/2, 2/2, 2/3 and said, "I kept on doing this until I could not write any more. What I ended up with was 120."

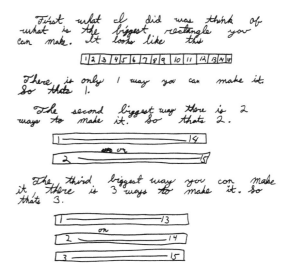

Figure 3.4. Rectangles are labeled with the positions of te squares at the ends.

Jared is one of several students who used the strategy of "drawing peaks" to count the rectangles (see fig. 3.5). His diagram became so involved that he stopped counting after 1 × 11 rectangles. It is interesting to note that by the third iteration, he had made three errors, but by the seventh iteration, other errors had compensated for the first three.

Kim and Park, from Seoul, Korea, presented the problem with a drawing in which all the individual boxes were rectangles, thus avoiding the question of whether a square is a rectangle. Their students had previously done this month's featured problem, "Counting Squares," and they seemed to connect their work on the "How Many Rectangles?" problem to that experience. Several students calculated the final sum by listing the numbers to be added as a string of vertical sums:

$$\begin{array}{r} 1\ \ 2\ \ 3\ \ldots\ 14\ \ 15 \\ +\ 15\ \ 14\ \ 13\ \ldots\ \ 2\ \ \ 1 \end{array}$$

The string yields 15 sixteens. Then, since this total counts every number twice, it must be divided by 2. Thus, the final total can be written as (16 × 15) / 2 = 120.

A group of teachers at Mayfield School in Middletown, Ohio, worked on the problem at an in-service meeting conducted by Catherine Mulligan of Fenwick High School. The teachers decided to present the problem to their classes from grade 1 through grade 5. The teachers differed in how they presented it. Some teachers explained the relationship between a square and a rectangle before the students began to work. Other teachers modeled counting the first few rectangles. Mulligan stated, "The results seemed to indicate that the children were unfamiliar with counting in situations where there is 'more than meets the eye.' Children were not accustomed to explaining their thinking in writing. Many of the dynamics of group problem solving were encountered."

One facet that makes this problem so rich is that depending on the approach or on the interpretation, several good questions are actually *embedded* in the problem. The first one came up again and again: "Is a square a rectangle?" Other questions are "Can two different rectangles be formed that share a common square? Two squares?" and so on. Others include "Do you see a pattern in how rectangles are being generated?" and "Have I counted or accounted for all the rectangles?" Donna Donnell, a second-grade teacher at Northwood Elementary School in Northwood, New Hampshire, posed this question: "Is there a rule for generating how many rectangles can be formed by a row of thirty squares? One hundred squares? *N* squares?" All these questions make this problem challenging for learners at many grade levels.

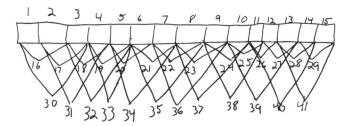

Figure 3.5. Three iterations of the "drawing peaks" strategy

It appears that all the students who attempted to solve the problem achieved a measure of success, because they were involved in a rich exploration that involved conjecturing and testing their conjectures. We would caution having the teacher model an easier version of the problem before the students work on it; that tactic can indicate a recommended approach and might limit the strategies that students would otherwise devise.

What Are the Students Telling Us?

In solving this problem, students are telling us that there are many ways that the rectangles can be counted. There are also many different models that can be created to illustrate the rectangles. Several students are telling us that they have not considered the relationship between squares and rectangles. They are not aware that squares are rectangles, with an added characteristic—equal side lengths. Several teachers who have tried this problem since it was first published in September 1997 said that before their class did this problem, they taught that squares and rectangles were different. Now they talk about squares as a type of rectangle. Another important thing that students are telling us is that they are able to successfully solve rich, challenging problems like this one.

4
DOUBLE THAT DOUGH

The Problem

Doubling Doughby created a mixture of dough that when placed in a warm oven doubles in volume every minute. He designed a pan, called a Doughby Pan, to hold the dough. He noticed that when a Doughby Pan full of dough was placed in the oven, it would exactly fill the oven in one hour.

Doubling Doughby usually had two ovens into which he placed a Doughby Pan full of dough, but one day one of his ovens was not working. He then had to place the two Doughby Pans full of dough into one oven. How long will it take for two Doughby Pans full of dough to fill the single oven?

Range of Possible Teacher Questions

- Do you see a pattern?
- Can you create a model to show what the dough is doing?
- Can you create a table?
- How much time was saved at the beginning by putting the two pans in one oven?

Where's the Math?

The math in this problem situation takes several forms. Students are engaged in problem solving and communicating mathematically throughout the exploration as they discuss the problem and figure out strategies for solving it. They are also creating models as they come up with ways to show how the measures are doubling. As they create tables to describe what is happening to the dough, they are representing and analyzing data. Students are also exploring exponential growth in a concrete, real-life situation.

SAMPLES OF HOW THE PROBLEM WAS SOLVED

It is clear that a problem like Doubling Doughby is a difficult one for children. Not only do children have a difficult time with questions that do not contain numbers with which to work, but several children tried to factor in information from their lives that was not pertinent to the solution. Many attempts did not produce acceptable solutions and appeared to follow the same reasoning. However, we are able to share two correct, well-presented solutions.

Mary Ferns, a preservice elementary education student, posed this problem to a fourth-grade class in Rock Island, Illinois. During a unit on number sense and place value, she used the problem as homework and gave students four days to bring in their answers. She also changed the time to cook from one hour to ten minutes. One student obtained the correct solution by making the following tables, which were part of his explanation.

Raw dough = 1 cup = 1 pan

1 minute	$1 \times 2 =$	2 cups
2 minutes	$2 \times 2 =$	4 cups
3 minutes	$4 \times 2 =$	8 cups
4 minutes	$8 \times 2 =$	16 cups
5 minutes	$16 \times 2 =$	32 cups
6 minutes	$32 \times 2 =$	64 cups
7 minutes	$64 \times 2 =$	128 cups
8 minutes	$128 \times 2 =$	256 cups
9 minutes	$256 \times 2 =$	512 cups
10 minutes	$512 \times 2 =$	1024 cups

Pan 1 = 1 cup

1 minute	$1 \times 2 =$	2 cups
2 minutes	$2 \times 2 =$	4 cups
3 minutes	$4 \times 2 =$	8 cups
4 minutes	$8 \times 2 =$	16 cups
5 minutes	$16 \times 2 =$	32 cups
6 minutes	$32 \times 2 =$	64 cups
7 minutes	$64 \times 2 =$	128 cups
8 minutes	$128 \times 2 =$	256 cups
9 minutes	$256 \times 2 =$	512 cups

Pan 2 = 1 cup

1 minute	$1 \times 2 =$	2 cups
2 minutes	$2 \times 2 =$	4 cups
3 minutes	$4 \times 2 =$	8 cups
4 minutes	$8 \times 2 =$	16 cups
5 minutes	$16 \times 2 =$	32 cups
6 minutes	$32 \times 2 =$	64 cups
7 minutes	$64 \times 2 =$	128 cups
8 minutes	$128 \times 2 =$	256 cups
9 minutes	$256 \times 2 =$	512 cups

"With the two pans of dough, the two amounts of dough at 9 minutes equals 1024, which is what one oven would hold. So the answer is 9 minutes."

Using patterns, similar to those given by the students in Fern's group, might make this task easier for students. Examining the data in a different manner, we could write the information shown in table 1 to compare using two pans with using one.

Even with these data, students may not be able to make any connections to the amount of dough required to fill the oven. A student who wants to continue this pattern may be disappointed, because very few calculators have a sufficiently large display to continue this pattern to 60 minutes.

However, we can see that the patterns for both the one-pan and two-pan expansion contain the same numbers. The number of pans for two minutes starting with one pan is the same as the number of pans for 1 minute starting with two pans. Likewise, the number of pans for 8 minutes starting with one pan is the same number of pans for 7 minutes starting with two pans. At this point, a correct conjecture is that placing two pans in the oven will fill the oven in 59 minutes, or one minute less than it takes to fill the oven starting with one pan.

Ferns's other student responses were incorrect but were similar to those reported by others. For example, six students thought that the answer was 10, two thought that the answer was 15, and one student thought that the answer was 30. She observed that "the students thought of this more as a cooking problem, and several related it to the use of a microwave or to other cooking experiences they have had. Students who had answered 15 or 20 minutes seemed to relate the problem to there being more dough, [so] it should take more time."

Julie Travis, a sixth-grade teacher in Spokane, Washington, shared work that she and one student had done. Travis commented, "In posing the 'Double That Dough' problem to one sixth-grade student, I orally explained the problem and asked if he might want to work on it. His initial response was 'That's easy, it's 30 minutes.' I replied, 'I think the problem is a little more complicated than that.' I then set off to try the problem myself. I used the same procedure he used later, in terms of doubling. Eventually, after carefully working through all those powers of two, I came to the end. That's when I wanted to say to myself, 'This was an easy problem had I spent more time thinking of what doubling really means.'"

In the work submitted, the student wrote, "First I thought that the answer would be 30 minutes because 30 minutes is half an hour. Then I talked to my teacher and she said, 'I think it's a little more complicated than that.' So I thought about it that night and figured out an equation—start at one and double 60 times for the 60 minutes in the hour, then divide by two because there are two pans in the oven instead of one; and that will be your answer." He proceeded to generate a list with sixty entries showing all the doubling he had done, including an eighteen-digit number for a solution.

Linda Brinkman, a fifth-grade teacher in Erie, Illinois, posed the original problem to her class, with students working in groups of three or four. Two groups indicated the correct solution of 59 minutes, although they did not show their work. One group attempted to make a list of the number of pans in the oven and created the following list of values before giving up: 2, 4, 8, 16, 32, 64, 128, 256, 512, 1 024, 2 048, 4 096, 8 182, 16 384, 32 768. Their calculations were correct, but they were able neither to complete their thoughts nor to see how to use this information to solve the problem. Five groups gave the solution of one-half hour (30 minutes), the time being simply cut in half. Written explanations were lacking from all groups.

Phyllis Laliberte's fourth-grade students in Middlebury, Vermont, also provided solutions. Thirty minutes was the solution attained by all but one student, who said that it was two hours. These students were unable to solve the problem, but they did articulate their solutions and reasoning well, as shown by a few student responses: "I figured it out by adding one hour plus one hour and that is two hours. Because if it takes one hour for one batch it takes two hours for two batches"; "It should take less time because there is double the dough, it should take half the time. Which is half an hour"; "It takes a half an hour. Because if one takes a half an hour to fill half of an oven. Half plus half equals a [w]hole, so it takes half an hour to fill the oven."

One of Laliberte's students posed, modeled, and correctly solved the following extension he had written:

What if Doubling Doughby had bought five brand new ovens that could heat two pieces of bread in ten minutes! How many pieces of bread could he heat in five hours?

What Are the Students Telling Us?

What can we learn from the responses to this problem? Without teacher guidance, students have a hard time starting this problem beyond giving the solution as half the original time. We are reminded of the importance of the first step in problem solving—understanding the problem. This problem is difficult for children because very little numerical information is provided. Although children may be able to examine patterns that have been constructed for them, they have a more difficult time organizing their thoughts to determine the need for, or to generate, a pattern from which to draw conclusions. Reducing the amount of time to ten minutes made the problem more accessible. It would be interesting to examine whether students who correctly solve the problem when given a 10-minute time frame can generalize the solution of 59 minutes for the problem as posed.

This is a problem for which students need to gather their own data. Teachers can build on this strategy to expand students' ability to solve such problems. Also, we must work to provide opportunities for students to reflect on their work, so that they confront data that will challenge their original, incorrect, thinking.

As a result of many experiences where this problem was done with children and with adults, it is clear that this is not an intuitive problem. As mentioned above, not only do children have a difficult time with problem situations that don't generate numbers, but because of the lack of numbers modeling becomes a problem as well. This is a problem that can be done very successfully with spread sheets, where the numbers that are input are the number of Doughby measures.

5
HOW MUCH FILM?

The Problem

South Main School has 618 students. Each student will have his or her picture taken next week by a photographer hired by the school. The photographer uses rolls of film having twenty-four exposures each. How many rolls of film will the photographer have to buy?

Range of Possible Teacher Questions

- Can you create a model to show how the problem can be solved?
- Are there any other ways you could represent the problem?
- Were there any "left-overs?"
- What does the left-over part mean?
- How did others around you represent or solve the problem?

Where's the Math?

There is a range of math in this problem situation. Students are engaged in problem solving and communicating mathematically throughout the exploration as they discuss the problem and figure out strategies for solving it. They are also creating models as they come up with ways to show how the rolls of film represent numbers of students. Students are also exploring the relationship between division and multiplication in a concrete, real-life situation. Some students solved the problem using repeated addition as well. Students also used estimation in solving the problem. One of the best parts of this problem is how students had the opportunity to deal with remainders in a real-life situation. They came up with very good ways to address the issue of the remainder.

SAMPLES OF HOW THE PROBLEM WAS SOLVED

THIS problem is a new take on an old "picture." The Third National Assessment of Educational Progress used the bus problem: "An army bus holds 36 soldiers. If 1128 soldiers are being bused to their training site, how many buses are needed?" (NAEP 1983). Because only 24 percent of a national sample of thirteen-year-olds solved the problem correctly, we have become more aware of the need to emphasize problems that involve division with remainders, especially when a need exists to augment the quotient for the solution to the problem.

The process of reasoning and looking back at the problem to determine the solution is "not an optional activity" according to Silver (1993), who stresses the fact that much more than correct computation is necessary for problem solving. An important question here is, What are some of the different ways that students approach this problem, whether or not they have been introduced to the division algorithm?

Beth Tagtmeir posed a brownie problem to her fourth-grade students in Moline, Illinois: "Mary has 100 brownies. She will put them into containers that hold exactly 40 brownies each." Tagtmeir guided her students' thinking by asking three questions:

1. How many containers can Mary fill?
2. How many containers will Mary use for all the brownies?
3. After Mary fills as many containers as she can, how many brownies will be left over?

These questions follow the format that Silver used in his research.

Most of the students reasoned through the problem in this way: 40 + 40 = 80, and that amount fills two containers, so you need one more container for the other 20 brownies. Jackie wrote, "She can fill three containers because 40 + 40 is 80, and that's not more than 100, and 80 is two containers; therefore, you need 3 containers to fill 100 brownies." Jessica used a similar strategy, but she chose pictures of containers to help keep track of the brownies. Her work is shown in figure 5.1.

Robin Ittigson at John Minadeo Elementary School in Pittsburgh, Pennsylvania, asks her fourth-grade students to write problems in their diaries along with explanations of their solutions. They had been using the strategy of making an organized table to solve problems but had not yet worked on division

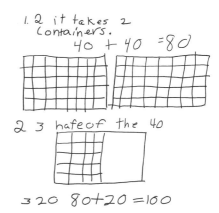

Figure 5.1. Jessica's work on the brownie problem

when Ittigson posed the film problem. She saw students draw on their prior knowledge as well as information they had learned about making a table, patterns, and multiples. Ittigson reported that her students were enthused about doing a "difficult" problem. She showed her camera to the students, and together they talked about whether one roll of film would be enough for all twenty-four students in the class.

Ittigson offered this account of her method of assessment: "I would assess students' work on the basis of the soundness of their strategy, whether they explained how they chose to solve the problem, and whether their approach makes sense with what is posed in the story. Problems arose around what to do with the remainder. I told them that this information had to be part of their solution."

One student in the class wrote, "I solved this by making a table, and on the top I wrote, 'Students,' and on the bottom I wrote, 'Rolls.'" (See fig. 5.2.)

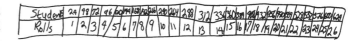

Figure 5.2. A student's table for the film problem

Another student said, "I solved this [problem] by making a picture graph and going up until I got to 618 or a little bit higher." (See fig. 5.3.)

Another student insightfully wrote, "You will need about 26 rolls of film. I solved this [problem] by adding up 24 until I got the closest. I could get to 618 so everyone could have their picture taken. He could

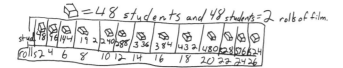

Figure 5.3. A creative picture graph shows this student's thinking.

take 6 more pictures with the leftover film. I could have doubled the 24s and added to make it a little easier to solve."

Terri Goyin of Belgrade, Montana, reported, "This was hard for fourth graders, but some began grouping the numbers in 100s—a usable number for them. Some put the students in groups of 25 and grouped those to 600. Then they added one more roll of film for the extra student in each group of twenty-five and then one more roll for the eighteen left over." (See fig. 5.4.)

Nadine Peirce posed the film problem to a group of her first-grade students in Redwood Falls, Minnesota. Ilene Hokanson, Title I lead teacher and Southwest Minnesota Coordinator for Cognitively Guided Instruction (CGI), wrote, "For the past four years Nadine has used the math teaching philosophy of CGI. Her children do problem solving every day rather than the traditional number crunching with 'naked numbers.' " Because of the age level of her students, Peirce gave them a choice of number sizes—143 or 618 students. Eight of her 24 students successfully solved the problem using 618 students, and one Title I student solved the problem using 143 students.

Some of the solution strategies that the students used, as described by Hokanson, follow.

"Michael and Kevin built 618 with 61 base-ten rods and 8 ones. They decided to change 4 rods in for ones and then combined pieces in different ways to make groups of 24. The students ended up with 25 groups of 24 and one group of 18 ... meaning 26 rolls were needed."

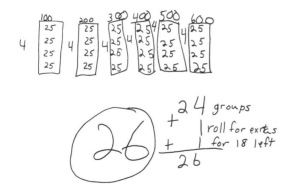

Figure 5.4. This student worked with more comfortable numbers.

"Kristen and Jackie made 618 of six hundreds blocks, 1 ten block, and 8 ones. They then marked every 24th block and counted their marks and got 26 rolls of film needed."

From the student papers that were sent to us, it is evident that problems such as these cause difficulties for students who approach the problem algorithmically without reflecting on the context of the problem. The following fifth- and sixth-grade responses give us some insights related to the overreliance on algorithms.

Many fifth-grade students at Hillview Elementary School in Lancaster, New York, approached this problem using the division algorithm, recording their answer as 25 remainder 18. One student wrote after recording such an answer, "I think this [problem] was easy because I knew right away what I was doing." Another student with the same answer wrote, "It was okay, but I probably got it wrong because I'm not good at division."

Mike Schoebroek, who teaches at Sherrard Junior High School in Sherrard, Illinois, gave the film problem to 107 seventh-grade students and instructed them to show a picture of the division or multiplication situation. Forty of the students solved the problem correctly.

Diane Devine, mathematics lead teacher, and Peggy Mar, sixth-grade teacher, from Williams Middle School in Chelsea, Massachusetts, posed the film problem to their students. They observed, "Although much time is spent in discussion of how estimation strategies and models can be used to problem solve, very few students in this class of thirty used the techniques. It is a real struggle to persuade the average students of this age group to spend more time thinking about a problem and using a variety of ways to find the answer."

The issue of the remainder opened up a discussion among these students on the necessity of the photographer's using mathematics so that enough film would be available to complete the job in one day. This real-life connection emphasized the importance of mathematics in performing well in one's career. Devine and Mar wrote, "It helps when problem solving also presents the teacher with a real-life connection to values that are very important to our job as educators of the whole child, not just the mathematical part."

Devine and Mar brought to our attention a concern in many classrooms. "Owing to the fact that many of our students speak another language at home, we find that another issue arises: How do we help students to explain problem solutions in words or create problems when English is not their primary language." They, as other teachers, are interested in learning techniques from teachers in similar environments.

What Are the Students Telling Us?

We learned so much from seeing how students solved this problem. They're telling us that they don't need to know division to explore division-related problems. Students were able to successfully use multiplication and addition to solve the problem.

Students showed us that they are able to use good estimation skills to solve the problem successfully. They also showed us that they can solve the problem using a range of numbers.

Students also told us that they are still developing understanding of the division algorithm. The situation proved much more challenging to those who used the division algorithm than to those who didn't.

Finally, students told us through their work that having a real-life connection helped them reason and strategize about the problem.

References

Silver, Edward A., Lora J. Shapiro, and Adam Deutsch. "Sense Making and the Solution of Division Problems Involving Remainders: An Examination of Middle School Students' Solution Processes and Their Interpretations of Solutions." *Journal for Research in Mathematics* Education 24 (March 1993): 117–35.

6

WHEN WILL WE REACH 1/2?

The Problem

Working with a hundreds chart, how many numbers can be covered that contain *only* the digit 1? (See figure 6.1.) If we used the digits 1 and 2, how many numbers could be covered that contain only the digits 1 or 2 or both? By following this pattern, how many digits are needed before the hundreds chart is half covered?

1	2	3	4	5
11	12	13	14	15
21	22	23	24	25

Fig. 6.1. Two numbers can be covered that contain *only* the digit 1.

Possible Teacher Questions

- What conjectures do you have about when we'll reach 1/2?
- Is there anything unclear about the problem? If so talk with others to see if you can clarify it.
- What patterns are emerging?

Where's the Math?

When students are engaged in exploring patterns and describing them, they are exploring algebra and number concepts. With this problem students have an opportunity do some reasoning as they test the common misconception that covering numbers using half of the numerals ought to yield a covering of half of the grid. As students explore and discuss their conjectures they are also communicating mathematically.

SAMPLES OF HOW THE PROBLEM WAS SOLVED

THIS problem is one that can be solved in a variety of ways. As a result, there isn't a need to be thinking along "one track" in order to achieve success in understanding and attempting to solve the problem. For example, those who prefer to use a hundreds chart to look for and eliminate numbers as they are formed can do so. Others who prefer a listing method can easily develop an approach for listing and organizing their thoughts. Unlike the problem in How Many Rectangles? (*Teaching Children Mathematics*, March 1997) organizational skills do not need to be too well-developed in order for a child to begin to see a pattern.

The "When will we reach 1/2?" problem is also a good one for children at a variety of developmental levels to solve. When shading in the number chart, a child who does not yet multiply need not see the array formed by consecutive numbers being multiplied in order to solve the problem. Angela Andrews' kindergarten class in Napierville, Illinois, is a perfect example. Students were practicing writing numbers first using only a 1's digit, then using 1 and 2, and so on. Ms. Andrews, the teacher who first shared this problem with us posed the problem above to the children. They noticed that there was an ⌐-shaped portion being shaded in on each successive day. The new ⌐ wrapped around the l that was there before. They were able to make predictions and test their predictions over a few days of adding a new digit to the list of digits that they could use (see fig. 6.2.).

Day 1	Day 2	Day 3	Day 4						
1	2	3	4	5	6	7	8	9	10
11	12	13	14	15	16	17	18	19	20
21	22	23	24	25	26	27	28	29	30
31	32	33	34	35	36	37	38	39	40
41	42	43	44	45	46	47	48	49	50
51	52	53	54	55	56	57	58	59	60
61	62	63	64	65	66	67	68	69	70
71	72	73	74	75	76	77	78	79	80
81	82	83	84	85	86	87	88	89	90
91	92	93	94	95	96	97	98	99	100

Figure 6.2. Many students observed this ⌐-shaped pattern.

Several students in Barbara Brannon's third grade class at Doss Elementary in Austin, Texas, looked at the problem as a successive multiplication problem, because the shaded portion was first a 1×2 rectangle that consisted of two numbers, then a 2×3 rectangle that consisted of six numbers, then a 3×4 rectangle that consisted of 12 numbers, etc. Some recognized a pattern by the time they added the 3 to the list of digits that they could use. They knew that in order to get beyond 50, they would have to choose numbers which would yield the first product larger than 50. Thus they chose a 7×8 which would be created when the digit 7 was included.

Ms. Brannon's class was distracted by two issues. The first was that they thought they were to create their own patterns. They were also bothered by their sense that the problem was asking for when the hundreds chart was to be exactly half-full. They were not sure that they could use 56 since it would fill more than half of the chart. This particular issue is a great one for students to discuss and attempt to agree on. After she noticed that students were bothered by these facts, Ms. Brannon created a Problem Sheet to focus students on the use of the digits. She asked the following three questions to focus them on the use of the digits.

- "Working with a hundreds chart, how many numbers can we cover that contain *only* the digit 1?
- If we used the digits 1 and 2, how many numbers could be covered that contain *only* the digits 1 or 2 or both?
- By following this pattern, how many digits are needed before the hundreds chart is half covered?"

As some good follow-up questions on the sheet, Ms. Brannon asked the children to share their thinking.

- "Describe the steps you took to find the solution.
- What did you think about or do first?
- What did you do next?
- Did you use a pattern to find a solution? If so, describe the pattern that you used.
- What did you learn from this assignment?"

This approach to working out the confusion without taking away the problem allowed for this to be a successful situation for students even when they did not first recognize the nature of the problem. If she had stepped in and begun to explain what they should do, then those who were struggling would have missed the opportunity to sort out their own confusion.

One child in Ms. Brannon's class, Quint, explained in response to the last question that "You can make patterns with questions that sound like the numbers would be scattered." He had envisioned the problem as one of finding numbers scattered about the number chart and eliminating the numbers. The benefit to him of solving this problem was that one could expect patterns in places where they were not anticipated.

A Variety of Solutions

This is an example of a problem that while it has a unique solution, there are a variety of solution strategies. Described here are some of the visual and some of the numerical strategies.

Ms. Andrew's class found the half-way point by using different colors to represent the new ⌋ that was able to be filled in on each successive day. Several of the third graders also used the ⌋ to shade the next level that could be written using another digit (see fig. 6.2.).

The organizational strategy for listing does not need to be sophisticated. As a result of listing how many numbers can be written at each level, a multiplication approach to solving the problem tends to be more visible. In each of the instances below, the last digit added to the list is a factor of the total of all of the numbers that can be written using those digits. The other factor of the number is the digit after the last digit written.

Digits Used	Number of Numbers Generated
1	2
1, 2	6
1, 2, 3	12
1, 2, 3, 4	20

Some students used the shading of the array to determine that a 1 × 2 was formed. That was followed by a 2 × 3, and then a 3 × 4, etc. The product of each of those (the area of the rectangle in squares) was equal to how many numbers were shaded in. The next step for them was to determine when the product would go over 50, which was using the digits 1 through 7.

One teacher said that her students guessed that there would just be two more each time. She suggested that they explore whether the trend continued if they tested this belief. They quickly found that this was not the case. It is the case that first two are added, then four, then six, then eight, and so forth. David R. in Ms. Brannon's class said, "At each digit, 2 more numbers were (added) than were added at the previous digit: 2 + 4 + 6 + ... when we reached the seventh digit we had a total of 56." It would be interesting, as a next step, to explore with children why it happens that there are two more added to the number being added. They could use the shaded portions on their numbers chart to generate ideas about this.

Another pattern that was recognized by several students was that of multiplying the last digit added to the list by the next consecutive digit. That would yield the total of the numbers that could be written using that many digits. For example, Zehaia, also in Ms. Brannon's class, said, "You take the highest number used times the next highest number which equals how many to fill in. Like if you had only numbers 1, 2, and 3, you would take the 3 and multiply it by 4 which equals 12. So there should be 12 squares to fill in. Numbers that have digits 1–7, there should be 56 numbers."

Students and teachers seemed to enjoy this problem. A final comment by Barbara Brannon was "This was such a good learning experience for us. We look forward to the next problem solver."

7 DECORATION DELIGHT

The Problem

The owner of a greeting card store wanted to decorate the front window with pictures of snowmen. Her design was finished when she decided to add color to the snowmen. She wanted to use four colors: one color for each of the hat, head, middle section, and bottom section of the snowmen. In how many different ways can the snowmen be colored for the window display?

Range of Possible Teacher Questions
- How many snowmen do you think you'll be able to make?
- Can you organize the data?
- Is there a way to pattern the change in color from one snowman to the next?

Where's the Math?

The math in this problem situation begins when the children estimate how many different snowmen they can make with the colors given. Students are engaged in problem solving and communicating mathematically throughout the exploration as they discuss the problem and figure out strategies for solving it. They are also looking for patterns in order to represent all of the different color combinations. They are representing and analyzing data as they determine how many ways the snowmen can be represented.

SAMPLES OF HOW THE PROBLEM WAS SOLVED

The feedback to "Decoration Delight" supplied many insightful looks at student thinking and classroom planning. We heard from teachers of pre–K through grade 6, which indicates that children of various ages can work on a similar problem.

Shelia Carr tried "Decoration Delight" with pre–K students. After modifying the original directions, she presented her version as a whole-class activity. She thought that it was hard for her students to determine the main purpose of the activity. As she progressed, students moved from errors and guesses to more correct solutions. She did a hat of one color at a time and noticed that the students "did catch on to the fact that six hats were used for each of the four colors." As might be suspected, younger children do not readily see similarities and differences. Ivey Parker's (Frank Long Elementary School, Hinesville, Ga.) kindergarten students "did not understand how their snowmen would be different if [their hats] were the same color." This lack of understanding led to a discussion that people of the same skin color were different, which helped the students better understand the problem. Students of several ages did the same thing as the students in Deborah Burnette's (Button Gwinnett Elementary School, Hinesville, Ga.) first-grade class. When given eight snowmen to color with the same color hat, each group used all eight snowmen, thus making duplicates.

After reading *A Three Hat Day* (Geringer 1985) and exploring the six ways in which three colored hats could be arranged, Linda Oliver (East Broad Street Elementary School, Savannah, Ga.) posed "Decoration Delight" to her third-grade students. Oliver commented as follows:

> I observed one group that started by having each member color a hat blue. Then they systematically colored the faces something different. But one member noted that for each blue hat and red face, there were two ways to color the remaining middle and bottom sections. Once this discovery was made, it was short work to conclude that for every hat of a specified color, there would be six ways to arrange the remaining colors. It seems that several groups came to this understanding at almost the same time.... At this time, we returned to the board and made some connections between the hat story and the snowmen display. I covered the hat on a snowman and asked them how many sections were left to be colored. Of course they answered three. Shekira and Jake noted that the three sections were just like the three hats—they could be six different ways. Casey noted that there are four hat colors, so you can multiply the four colors times the six possible arrangements to find that there are twenty-four in all.

Oliver made table 7.1 and asked for any patterns the students could see. The students predicted that five colors could be arranged in 24×5, or 120, ways. Oliver noted, "They marveled over how many options were available. They noted with amazement how many more ways six hats could be arranged, and, of course, there was no stopping us now! We had to know how many ways we could arrange seven hats, and eight hats, and so on, up to ten hats."

TABLE 7.1

Number of Hats	Number of Ways to Color
1	1
2	$2 = 2 \times 1$
3	$6 = 3 \times 2$
4	$24 = 4 \times 6$

Many teachers cited other examples of good reasoning. Cindy Bosela's first-grade students, after spending thirty minutes and generating twenty-five combinations, were asked how to check their solution. "One student said that we could put all the snowmen with black hats together, which we did.... Another said to put all green hats together, and they began to see a pattern. One said that he thought they would all have six in their group, since the first two had six." When they found that the next color had seven, they reexamined their combinations and located the duplicates.

The notion of controlling one color at a time—whether the top or the bottom—came up repeatedly from students at various grade levels. Fifth-grade students in Jennifer Young's class extended this idea and concluded that "only two body parts of the snowmen could be the same color in a group." For example, only two snowmen could have green hats and purple bottoms. A student in Audrey Hayes's (Joseph Martin Elementary School, Hinesville, Ga.) class used similar logic and noticed that "if two snowmen had the same bottom two pieces, then their top pieces were switched.... So she generalized this to say that every snowman had a partner who had the same bottom two pieces, but switched top pieces and so they were 'sort of like twins, but their head and hat colors are switched so people can tell the twins apart.'"

In describing her fourth-grade-students' work, Varina Moser made this statement:

My students really enjoyed completing this activity. I enjoyed watching them work out their different ideas. They were simply amazing.... They were discussing how many ways and debating their ideas about how many combinations could be found.... Many of them thought that because I still had a rather large stack of blank snowmen, then there must be more ways to color them. However, our student leader convinced them by showing [them a table like table 7.1] that we were finished.

Kristine Weaver (Klondike Elementary School, West Lafayette, Ind.) described some frustrations in getting first-grade students to discuss and reflect. She intervened when her students thought that the answer must be four because only four ways were possible to have different colored hats. The question she posed was, "Is it possible to have two *different* snowmen, both with a blue hat?" This same situation was described by Cathy Pascone (multiage first- and second-grade classroom, Turner Elementary School, Wilkinsburg, Pa.) and Bill Shidler (fifth grade, Klondike Elementary School). Shidler noted that some students used the digits 1, 2, 3, and 4 rather than colors to solve the problem in an orderly fashion. Several teachers from Klondike Elementary School—who regularly videotape their classes, reflect on the videotapes, and write in journals about their reflections—used this problem with their students to see how children across various grade levels approached it.

When Andrea Magnifico presented "Decoration Delight" to her first-grade students (C. J. Mangin Elementary School, Monaca, Pa.), someone mentioned that this problem seemed like another problem that they had completed in class:

If you have four letters, what are all the combinations of letters that can be made?

Although most students colored randomly, one pair of students remembered the method used to solve the four-letter problem and applied it to this problem (fig. 7.1). Kelsie explained, "I know I found all the ways because I switched the colors around. I did [the] problem with letters. That was almost the [same] thing." Magnifico stated, "What surprised me the most was that more students did not apply the strategy we used to solve the four-letter problem. I thought once they made this connection, the new problem would be very easy to solve." Jill Ferguson (Regal Elementary School, Spokane, Wash.) created a snowman template in Kids Pix Studio for her third-grade students. This abstract form did not work as well as planned, so she revised the problem to use only three body parts and three colors and suggested using multilinks to build the snowmen. She also showed the problem to a parent after school and spent "an hour and a half constructing models and gaining meaning while at the same time forming a very positive teacher-parent bridge."

Rhonda Schubert and Elizabeth Halsey-Sproul (second grade, Golden Hill Elementary School, Florida, N.Y.) "were disappointed that the students seemed to feel that the only way to solve the problem was to color the snowmen." A few days later, they presented a similar problem involving a cowboy with two different shirts, two different hats, and two different pants and asked the students how many different outfits he could wear to the rodeo. The students were unable to keep track of what they were finding and concluded that a list would be useful. "Within the allotted class time—forty-five minutes—most students were able to list a solution, and several had time to write an explanation."

Not only did Ward Moberg (Osceola Middle School, Osceola, Wis.) present this problem to forty-five fifth-grade students, but he also generated more than fifteen solution strategies and approaches from their work.

My students had no difficulty understanding the problem. I directed the students to work individually for five minutes to develop a strategy. Although they all had crayons and colored pencils at their disposal, only a few used them. The majority wrote "g" for green, and so forth. However, by the end of the class period, most were coloring. Several misunderstandings came to light when we discussed their first efforts to solve the problem.

Moberg indicated that students had difficulty explaining their reasoning in writing, so he let them discuss their ideas. Even from fifth graders, the reasons were mostly tied to trial and error. As a pencil-and-paper assessment, he gave a similar problem involving a daisy chain with three links, each link of a different color. He was pleased to find that the students employed strategies from the snowmen problem to solve the daisy-chain problem. After analyzing the students' work, Moberg stated, "Not many like Tyler [who solved the problem in five minutes] could see the pattern inherent in the problem.... I think Tyler is the only one who saw the structure of the problem and could reason that if six are of one color,

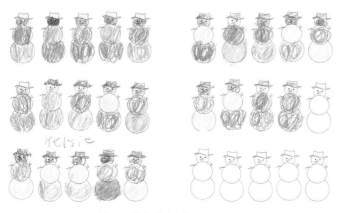

Figure 7.1. Kelsie's snowmen

the same will hold true for the remaining ones. Even those who organized by the same colored hat had to actually try all possibilities before they felt secure [of] their answer."

What Are the Students Telling Us?

What do we learn from the students' work? Students are powerful problem solvers who can use patterns and reasoning to determine solutions. Students of various grade and ability levels can solve problems of this nature. However, most students will use a trial-and-error strategy in the absence of guidance. Teachers are important because they must frame the question in a manner that allows students to take ownership. Teachers must decide how much direction they will offer and what questions they must ask to keep students focusing on the heart of the task instead of just finding another snowman.

We also see that some students need manipulatives, but others do not. Although students may have worked virtually identical problems, they may not see how the problems are the same, which indicates to teachers that they must continue to present problems that allow students to make these connections. Even though students have a difficult time reflecting on their work and expressing their ideas, teachers can still learn about students' thinking by examining their work on problem solving.

We thank all teachers who sent us material to read and consider. Space does not allow us to acknowledge each person, but we are most appreciative of your contributions. Please continue sending us student work.

Reference

Geringer, Laura. *A Three Hat Day*. Illus. New York: Harper & Row, 1985.

8
YOU GOTTA HAVE HEART—AND BLOOD!

The Problem

WITH this problem situation, rather than posing one problem for everyone to adapt and try with students, several pieces of information are presented that have the potential to spark exploration and questions on the part of a class at any grade level, K–6.

As a focus for exploration, have groups of students pose questions and write problems to be solved that are related to the information below. Then students can explore how to answer the questions and solve the problems generated. They can pose the problems for other groups in the class. Perhaps a health professional or Red Cross volunteer could visit your class and share his or her expertise. If you have access to the Internet, your students could collect information from many sources. Some sites are listed below.

We encourage you to pose problems for students as well as allowing them to pose their own problems and ask their own questions. Encourage students to write about the experience of posing and solving the problem(s).

In a recent ad from the American Red Cross, a very interesting fact was presented with an attached plea for donations.

> "Every year your heart pumps 2,625,000 pints of blood. Surely you can spare a few."

Donating blood has been a way of saving lives for a long time. The Red Cross took on its first nation-wide organized drive for blood in 1941. The American Red Cross has a Web site at www.crossnet.org/. This site contains very interesting and valuable information about the history of the Red Cross, including important blood donor information like the following.

> "In January 1941, the Red Cross blood donor project was organized at the request of the Surgeon General of the Army and Navy. During the four-year period beginning in February 1941, when the Blood Donor Service was established in New York City, six million Americans donated blood, most giving twice on average with nearly 1.5 million volunteers giving at least three times. Before the attack on Pearl Harbor (December 7, 1941), monthly blood donations averaged 2,735 pints. Immediately following the attack, Blood Donor Services expanded as donations increased, new centers opened, and the armed forces asked the Red Cross for greatly increased amounts of blood to be processed into dried plasma for the military. Within four months after the bombing, blood donations jumped to 53,770 pints."

> "During the Second World War, a total of 13.5 million units of blood were donated in the US, through the American Red Cross."

Another source of information on the Internet is the site of America's Blood Centers (ABC). ABC is a national network of nonprofit, independent community blood centers. It's members collect, test, process and deliver approximately 48 percent of the nation's blood supply to hospitals in 46 states and are fully licensed and regulated by the FDA.

According to America's Blood Centers, hundreds of high school students become blood donors each year, accounting for about 10 percent of the nation's blood supply. Recently, the Luann comic strip devoted the week of April 7, 1997, to a teen's decision to donate blood.

Other information, as gathered from the internet includes interesting facts about our nation's blood supply as presented at America's Blood Centers' web site, www.americasblood.org/:

> "Every three seconds someone needs blood"

> "The shelf-life for refrigerated red blood cells is 42 days. Platelets have only a five-day shelf-life. Frozen plasma can be stored for as long as one year."

> "Approximately 40,000 units of blood are used each day in the United States. One unit is the equivalent of one pint."

> "Because blood donations often are separated into several 'components' (red blood cells, plasma and platelets, for instance), one donation can help save three lives."

"Donors may give only one unit of blood at a time. Whole blood can be donated every 56 days, plasma every four weeks, and platelets 24 times a year."

The Franklin Institute Science Museum has a site describing the purpose of blood in our bodies at sln2.fi.edu/biosci/blood/types.html.

"The average adult has about five liters of blood living inside of their body, coursing through their vessels, delivering essential elements, and removing harmful wastes. Without blood, the human body would stop working."

Another interesting fact about our blood and its transportation system in our bodies was found at another site of the Franklin Institute Science Museum. sln2.fi.edu/biosci/vessels/vessels.html

"If you took all of the blood vessels out of an average child, and laid them out in one line, the line would be over 60,000 miles long! An adult's vessels would be closer to 100,000 miles long!"

Some other interesting sites are listed below, but are by no means a complete list.

www.cordblood.html
sln2.fi.edu/biosci/blood/blood.html
sln2.fi.edu/biosci/heart.html

Range of Possible Teacher Questions
- Which facts make for better problem situations?
- Do you see any patterns in the problems you're writing?
- Can you use estimation in the process of solving your problem?

Where's the Math?
There are several areas of mathematics that are touched on with this problem situation. Students are using problem solving in order to set up challenging problem situations for their peers to solve. They will be communicating mathematically when they are working together to write the problems and in discussing the problems their peers give them to solve. Students will also be making connections between math and other subject areas as they explore the mathematics involved.

SAMPLES OF HOW THE PROBLEM WAS SOLVED

IN RESPONSE to this challenge, several intriguing facts about the heart and about blood were shared with teachers. We asked you to share this information with your class, along with any other information the students might collect from the Internet (some interesting sites were provided) as well as from other sources.

Holly Weise of the American Boychoir School, in Princeton, New Jersey, shared all of her fifth grade students' problems with Problem Solvers. Included here are some of the problems that her students wrote, along with Holly's approach to presenting the problem. They were doing a unit on the circulatory system at the time "You Gotta Have Heart—and Blood" was sent out.

The assignment was as follows:

"General Directions:Write two problems that are based on any of the information given below. Then solve your problems. The facts to base your problems on appear below:

- Every year your heart pumps 2,625,000 pints of blood.
- The average adult has about five liters of blood living inside their body, coursing through their vessels, delivering essential elements, and removing harmful wastes. Without blood, the human body would stop working.
- Approximately 40,000 units of blood are used each day in the United States. One unit is equivalent to one pint.
- Donors may give only one unit of blood at a time. Whole blood can be donated every 56 days, plasma every four weeks, and platelets 24 times a year.
- Because blood donations often are separated into several "components" (red blood cells, plasma and platelets, for instance,) one donation can help save three lives.
- The shelf-life for refrigerated red blood cells is 42 days. Platelets have only a five-day shelf-life. Frozen plasma can be stored for as long as one year."

Ms. Weise's class posed many problems related to conversions. It appeared that they were interested in converting some of the information into forms that were more meaningful to them.

For example, Ned Milly asked the following question in figure 9.1.

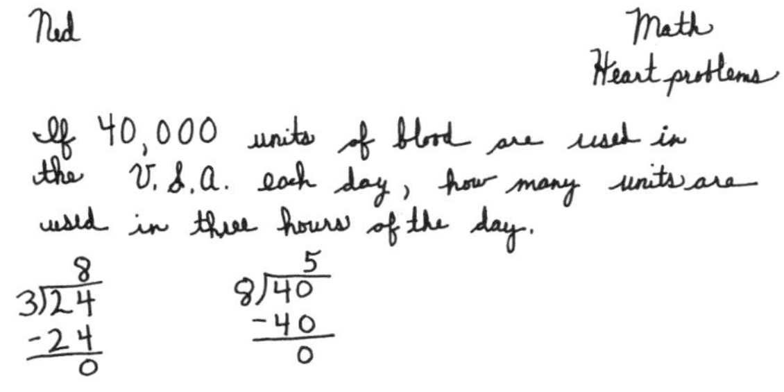
Ned

Math
Heart problems

If 40,000 units of blood are used in the U.S.A. each day, how many units are used in three hours of the day.

$3\overline{)24}$ = 8, −24, 0

$8\overline{)40}$ = 5, −40, 0

Figure 9.1. Ned extends the units-of-blood question.

Ned answered his own question by doing long division in an abbreviated form. He removed the factor 3 from 24, which left him with a factor that was common to both 24 and 40. Removing the factor of eight from 40 left Ned with a quotient of five. He added the zeros to the five to get the information that 5,000 units would be used in three hours. It would be interesting to hear from Ned whether there was a connection between choosing three hours as his time frame, since that left the eight three-hour time periods that yielded the round number of 5,000 units. Choosing a rate per hour would have left Ned with 166.67 units per hour. If Ned did intentionally select his time frame the way he did, the teacher has some insight into Ned's ability to manipulate numbers and data in order to report findings, based on the problem he wrote.

Another problem that several of the boys posed is represented in the work of Aaron Smyth.

"If you can give one pint of blood every 56 days, what is the maximum amount (number) of times you can give a pint each year?" That seems a reasonable question for a new donor. All students who posed this question appeared comfortable with the notion that one would give only six times, not 6.52 or 6 1/2 times.

Addison was curious about the rate per hour of blood being pumped by the heart. He posed his question the following way. "Every year your heart pumps 2,625,000 pints of blood. If there are 365 days in a year and 24 hours per day, how many cups does your heart pump in an hour?" (See figure 9.2.) Addison found out that the heart pumps approximately 598 cups of blood per hour. It would be interesting, as an extension, to model how much blood that is by trying to pour 10 cups of water in a minute or 100 cups of water in 10 minutes, and then discussing whether one would be able to keep up such a pace over time.

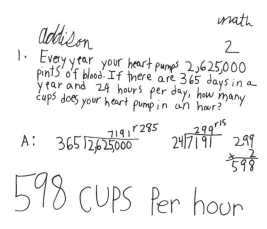

Figure 9.2. Addison extends the pints-of-blood-pumped problem.

A final question that was posed by T. J. DePaola, in Ms. Weise's class would be an interesting one to pose to a class. "The heart pumps 2,625,000 pints of blood a year. How many pints of blood would your heart pump after 22 years if you gave a pint of blood 5 times a year for five of those years?" This is a good question to pose in order to get the class thinking about how soon the blood is replenished, about how to account for a period of time where there is a slight aberration in the data, and about the overall effect of that aberration.

Another possible exploration that might be done as a result of studying the amount of blood pumped by the heart would be to allow groups of students to share their individual results with one another and then extend their questions further, or create new problems as a group. It would also be interesting for students to create a page for the book, *Counting on Frank* by Rod Clement, where the young man does some exploration based on the questions written by the group, or by individuals in the class. They would probably put their page in right before or after the one about the mosquito!

Thanks to Andy Reeves, Director of Editorial Services at *Teaching Children Mathematics*, and Danny Breidenbach, past journal editor for *Teaching Children Mathematics,* for the ad and the idea that sparked this "Problem Solvers."

References:

Rod Clement. *Counting on Frank*. Gareth Stevens Publishing: Milwaukee, Wisc.: Gareth Stevens Publishing, 1991

9
HOW MANY SANDWICHES?

The Problem

The Attribute family was going for a picnic. They made sandwiches from items that were in their refrigerator. They did not count all the sandwiches, but they did keep track of the various sandwiches they made. From the following information, can you help them determine how many sandwiches were made? A sandwich does not have more than one piece of any ingredient, although it may contain several different ingredients.

- Thirteen sandwiches had a slice of cheese.
- Fourteen had a slice of salami.
- Thirteen had a slice of tomato.
- Eight had a slice of cheese and a slice of tomato.
- Three had only a slice of salami.
- Five had a slice of tomato, a slice of cheese, and a slice of salami.
- Eight had a slice of tomato and a slice of salami.

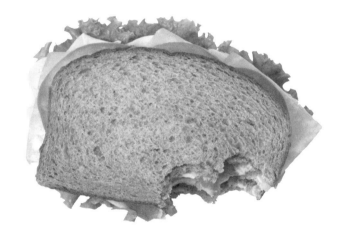

Range of Possible Teacher Questions
- Can you tell how many types of sandwich are possible?
- What strategies are you using to determine what goes on the sandwiches?
- Can you think of any ways to organize the information in order to keep track of it?
- Which pieces of information tell you exactly what was on a sandwich? Which don't?

Where's the Math?

There are several math concepts in this problem situation. Students are engaged in problem solving and communicating mathematically throughout the exploration as they discuss the problem and figure out strategies for solving it. They are also using logic and creating models as they come up with ways to account for the different combinations of ingredients in the sandwiches. Students are also looking at sets and subsets as they solve the problem.

SAMPLES OF HOW THE PROBLEM WAS SOLVED

THE responses to this problem showed that creative approaches were developed through collaborative group work and that the problem helped enrich a mathematics lesson. Sandra Kelly, the schoolwide enrichment teacher/coordinator at Spruce Run School in Clinton, New Jersey, wrote, "With a total inclusion philosophy and a schoolwide enrichment model, staff members are always looking for great ways to differentiate instruction to meet the needs of *all* of our students. The mathematics problem dealing with the sandwiches is the type of activity that can be used successfully in small math groups to differentiate/enrich math lessons."

Kelly met with second graders once a week for three weeks to work on the problem. She read it to the students and discussed with them the many ways to solve problems. They worked on the problem in groups and developed their own solution methods. George, Trevor, and Jimmy first approached the problem, as did other children of various ages, by adding the numbers: 13 + 14 + 13 + 8 + 3 + 5 + 8 = 64. However, after beginning with addition, George wrote, "but found out that was the wrong way" and then verbalized that some pieces had more than one ingredient. The group decided to make the sandwiches out of paper. They designed cheese, salami, and tomato on the computer and asked Kelly to design bread for their sandwiches. (See fig. 9.1.)

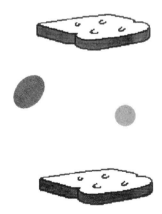

Figure 9.1. Second graders modeled sandwich ingredients to solve the problem.

The group cut out the ingredients and pasted the sandwiches together in rows on the floor. The students found twenty-four sandwiches rather than twenty-one because in realizing "that we would have to put the salami in a category all their own," they made six "salami" only sandwiches rather than three. They also made five sandwiches with cheese only and no salami-and-cheese sandwiches. Including the three salami-and-cheese sandwiches would have yielded three fewer sandwiches, giving them the twenty-one sandwiches they needed. This thinking was quite impressive for second graders who were working on a problem that challenged sixth graders. A considerable number of students found twenty-four sandwiches by making six "salami only" sandwiches rather than three. Many others simply added the numbers from the last four clues: 8 + 3 + 5 + 8 = 24.

In the larger number of responses from fourth-grade classrooms, the reasoning required to find the solution seems to be more developed, as exemplified in Blair, Becca, and Annie's response from Sheryl Orman's fourth-grade classroom in Saint Paul, Minnesota. They found their answer of twenty-one sandwiches "by noticing that you did not say that these were the only combinations of sandwiches."

Fourth graders in Heidi Vanderzee's class at Farwell Elementary School in Spokane, Washington, reflected similar thinking. One student wrote, "Then it said, 'Thirteen had a slice of tomato,' that doesn't mean you add 13 more—it means that the total is 13." Another student from Spokane wrote, "Next we put down 3 more salamis because we needed to do the logic and it said 8 had a slice of tomato and salami." The first student's thinking led to the solution of twenty-one, but the second student only found sixteen sandwiches.

Jean Stefanik from Amherst, Massachusetts, sent in a fourth-grade response written by Michael (see fig. 9.2). "I use the information to fill in as many spaces as I can.… After I fill in all the sections, I add them up. Then I have the answer."

Cathy Hayes from Te Puke, New Zealand, shared how the level of thinking of her year-6 class of nine- and ten-year-old children developed with her guidance. After reading the problem with the class and discussing different possibilities for fillings, students were told that they would have five days to work on the problem. Hayes wrote, "At this point I gave no help nor had any manipulatives for the children to work with. The immediate reaction of the children was that it was an easy problem, with most children selecting to add up all the categories. They took it at face value."

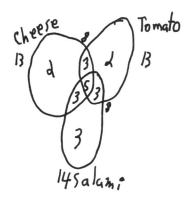

Figure 9.2. Michael's solution

On the second day, Hayes told the students that the thirteen cheese sandwiches need not be just cheese but could be cheese and tomato; cheese and salami; cheese, tomato, and salami; or just cheese. "This was a breakthrough piece of information for the children, as it opened up all the possibilities. The children then went off with great excitement to work on the problem, and they were engrossed." Although cubes were available for the students to use, they decided instead to use tally marks or symbols for the categories.

Some students reported their solutions and strategies on day 3, but no one had attained a correct answer. The students then discussed the categories of sandwich fillings, and they were able to see that the logic they had used was incorrect. On day 4, the discussion involved the three attributes, or fillings, and the topics of union and intersection. Two pairs of students solved the problem on day 4, and just over half of the class had solved the problem by the end of day 5.

Hayes noted that all the students who solved the problem used a variation of the same method: "Draw three circles and label. Put in the three salami and then the five salami, cheese, and tomato. Make up the figures to eight in the cheese/ tomato and tomato/salami sandwiches. Then it was just a matter of making numbers to meet the thirteen and fourteen totals."

When Tom Lewis posed the question to his fifth-grade students at Jane Addams School in Moline, Illinois, they took a few minutes to figure out how many different kinds of sandwiches were possible. "I suggested that this information might help them organize their charts and tables. They concluded that seven different sandwiches were possible." The class worked through the problem together using Venn diagrams to help them reason that they needed twenty-one sandwiches. Lewis wrote, "I thought the problem was over, we had had a good discussion, most of the kids were comfortable with the solution process, and one of the kids said that all we needed to do was take the seven combinations and multiply them by three and we had twenty-one. Everyone, of course, nodded and agreed with him. My questions for him were, 'Where did the 3 and 7 come from?' and 'Was the 3 × 7 = 21 part of a pattern for the solution or just a coincidence?'"

Brenda Deihl's sixth-grade students at Stroudsburg Middle School in Pennsylvania wrote about the processes that they used to solve the problem. The students' reasoning and communication about mathematical thinking demonstrate the kind of results that can be expected when students are asked "to demonstrate how you arrived at your answer, what strategies you used, what worked, and what didn't."

Jon used colored paper and cut out thirteen slices of cheese, fourteen slices of salami, and thirteen slices of tomato. "I put 8 pieces of cheese and salami on top of each other. Then, put aside 3 salami. I put 5 pieces of salami on 5 of the 8 cheese and tomato. Five of the 8 tomato and salami were already set out, so I set out 3 more salami with tomatoes. With the remaining 2 tomatoes and 2 pieces of cheese, I made them each into a new sandwich.... The final answer is 21 sandwiches."

Jared tried two approaches to the problem. His first attempt involved "making a sandwich mathematical tree." That strategy did not work for Jared. "Try number 2: This time I made a Venn diagram. As I went along, I checked that all the ingredients in groups of two and three equaled the amount of the original ingredients. I added all of the overlapping numbers that were in one of the circles and see if the answer equaled to the big part of the circle. Then I realized that I needed a third group of two ingredients, cheese and salami. I then added all of the remaining numbers together getting the answer of 21."

Molly clearly described how she approached the problem. "First try: In my first attempt to solve the attribute puzzle, I felt that pictures could give me a better understanding of the solution process I was undergoing. I drew pictures of slices of bread with spaces in between to add any ingredients. I started with thirteen sets of bread with cheese in the middle. Next, in eight out of the sandwiches I drew tomato slices. Next, I drew fourteen sandwiches with salami in the center. I then drew eight of the salami sandwiches with a slice of tomato. This was my mistake. I had already too many sandwiches with salami. The final step was circling the three 'just salami' sandwiches. After calculating my effort, I came to the conclusion that there were too many sandwiches with salami and I needed to try again. I decided to use a Venn diagram this time. This method seemed easier to organize and keep track of my steps." See figure 9.3 for Molly's Venn diagram with her notes explaining her thinking processes.

Meghan recorded her thought process with a color model and photographed each step toward her solution. She used twenty-eight red napkins that stood for sandwich bread; and construction-paper strips in black for the salami, yellow for the cheese, and green for the tomato. She wrote, "Rather than using endless amount of paper to keep track of progress, a camera and moveable manipulatives are better and more efficient choices. The problem is solved in steps, keeping organized and moving the strips accordingly." Figure 9.4 shows the final step in her solution.

Deihl's students appreciated the importance of communication in mathematics. As Christina wrote, "Even though I was unable to figure out an answer that works, I will show you what I did."

The richness of the sandwich problem is demonstrated by the way that teachers used it beyond the classroom. Claire Kelley from Plympton, Massachusetts, posed this problem at Dennett Elementary School's Family Math night. Rita Brown gave the problem to the Board of Education in Dell Rapids, South Dakota, as an example of the kind of problem solving that students need to do. The members of the board were very interested in the problem and possible solutions, so Lois Brown's sixth graders accepted the challenge of solving the problem.

In addition to the teachers already mentioned, we thank Janet Hazen, Seton Catholic Junior High School in Moline, Illinois; Colleen Hummed, Title I

Figure 9.4. Meghan's color-coded solution

teacher in Surrey, North Dakota; Gail Winter and the Belle Morris Math Club in Knoxville, Tennessee; Kim Markus-Jones, Stuart Burns Elementary School, Burns, Tennessee; and Mary Forrester, Sturgis Elementary School in Sturgis, South Dakota, whose students made chocolate-chip, M&M, and gum-drop cookies for the Attribute family, for sending in their solutions.

What Are the Students Telling Us?

There were several creative approaches developed by students through collaboration, as mentioned earlier. The children reminded us during exploration of this problem that good, challenging problem situations take time to think about. We also found that using models or manipulatives helped the problem solvers. Also, communication was very helpful in determining what's not told as well as what is told to the problem solver.

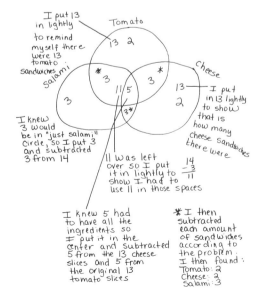

Figure 9.3. Molly's Venn diagram and explanation

10
SHUFFLING A LINE

The Problem

If you have twelve students in a row, as pictured, how many children must be "moved" so that the line of twelve alternates between boys and girls?

In this situation we will count a move when one person is taken out of line and inserted in another place. Other students may glide to the side to allow room for inserting a person in a new position. This glide does not count as a move.

Can you find a general pattern for the number of moves for any number of students in a line that has the same number of boys and girls and that has the girls originally at one end and the boys at the other?

Range of Possible Teacher Questions

- What is a move?
- Can you develop any "moving" strategies?
- Can you create a model to show how the children are moving?
- Do you see a pattern to describe how the students are moved?

Where's the Math?

The math in this problem situation takes several forms. Students are engaged in problem solving and communicating mathematically throughout the exploration as they discuss the problem and figure out strategies for solving it. They are also creating models as they come up with ways to show how to move the children. Students are using reasoning as they determine strategies for making the moves and as they look for moving patterns.

SAMPLES OF HOW THE PROBLEM WAS SOLVED

The responses to "Shuffling a Line" indicate that children can work with problems in which they have to decide what some words mean. In this problem, the definition of *a move* created an opportunity for such a discussion. However, even with the definition of a move clarified, many students were not able to arrange their moves to obtain the minimum number of moves. Of course, some of the alternative interpretations of a move generated interesting solutions.

Marianne DeRise (Kreiger School, Poughkeepsie, N.Y.), using six Unifix cubes of two different colors "had to demonstrate more than once what [she] was looking for." Several students obtained six moves with twelve students and eight moves with sixteen students, using a strategy of working "from the outside in" or "from the inside out." One third-grade student used the "outside in" strategy and referred to each move by the position in the line (see fig. 10.1). Note that his strategy leaves boys in both middle positions.

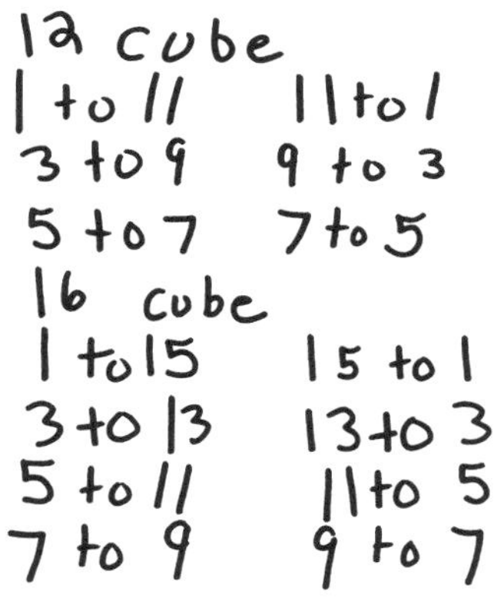

```
12 cube
1 to 11      11 to 1
3 to 9       9 to 3
5 to 7       7 to 5
16 cube
1 to 15      15 to 1
3 to 13      13 to 3
5 to 11      11 to 5
7 to 9       9 to 7
```

Figure 10.1. A third grader's "outside in" strategysaw.

Jan Hensley (Geneseo, Ill.) gave the problem to three children—a kindergartner, a fourth grader, and a sixth grader. The fourth-grade student first used eight moves in her solution and later used five moves. The sixth-grade student first used six moves to alternate people and then later found a solution in five moves by sliding the last boy over instead of actually moving him. The kindergarten student took six moves to alternate people and was somewhat quicker to solve the problem than the fourth-grade student.

Kelli Ekstrand (Buchanan Elementary, Davenport, Iowa) also gave this problem to three students, two in the third grade and one in the sixth grade. She chose three different manipulatives: Unifix cubes, paper squares, and M&M's. One third grader "didn't spend much time thinking about the problem. He could see a pattern right away. The student took a cube from one end and placed it between the two on the other end. He followed this process until his cubes were in an A-B pattern. It took him five moves." He used essentially the same pattern for each manipulative. The other third-grade student "began by taking cubes from both sides and intermixed the cubes to achieve an A-B pattern" in six moves. The first time she used paper squares, she also needed six moves; the second time, she was able to do it in five, as she did with the M&M's. When the problem was presented to the sixth-grade student, she had many questions and began moving the cubes in no organized fashion. In a creative fashion, she moved all the cubes of one color out of the line and inserted them into another place in the line, counting this action as one move. With the paper squares, she started in the middle, worked to the end, and completed the task in five moves. With M&M's, she moved the pieces from end to end and used six moves. It is interesting to note that the students did not always operate in the same way with the different manipulatives.

When Carol Drish (Ridgewood School, Rock Island, Ill.) used purple and blue paper squares to present the problem to her third-grade students, their understandings varied. One student thought that the purple and blue squares had to be exchanged. Two students thought that if they spread their fingers and held on to two squares, their action counted as one move. One used six moves to obtain BBPPBBPPBBPP and thought that he was done, and one stopped after six moves with no apparent pattern. Other students used different strategies and ended up with three, five, six, or seven moves. Drish reported that "most students didn't do a very good job of explaining their reasoning," so that it was difficult to understand how they worked the problem. However, she also found that "it was interesting to watch students who weren't good pencil-and-paper math students really try to solve this problem."

The students in Liz Bushman's (Lincoln Elementary School, Dixon, Ill.) third-grade class attempted the task by using cubes. Six of her students used six moves; two students, starting at one end and using a slide-and-replacement strategy, used five moves. Their strategy is shown in figure 2. Her other students had various misunderstandings about the task or its solution.

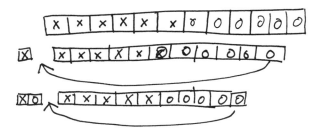

Figure 10.2. A third grader's "slide and replace" strategy

What Are the Students Telling Us?

For the work that these teachers have shared, we see that students of various grade levels can engage in the same problem and often have the same questions related to understanding the problem. For this problem, all students had to grapple with the definition of a move.

Also, the samples show that teachers need to help children reflect about their work. It is challenging to explain one's thinking. Students need experiences that allow them to develop expertise at explaining their thinking. One way to help students develop this skill is to allow them time to discuss what they are thinking as they solve problems. Another way is to give them time to reflect on the process of solving a problem after they have worked on it.

The students reinforced something for us in this problem that is important to remember. Some of the students showed inconsistencies in their understanding of the problem when the manipulatives they were using changed. This supports the idea that we construct our own knowledge in ways that are meaningful to us individually. Students need to use strategies and manipulatives that make sense to them.

11
COUNTING SQUARES

The Problem

The steps below are made of squares. How many can you count? What if we add another row at the bottom? How many squares could you count? Continue adding rows along the bottom. Can you find a pattern that describes how the total number of squares is growing?

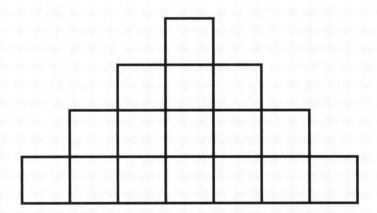

Range of Possible Teacher Questions
- Are all the squares the same size?
- Do you see any patterns?
- Have you found any hidden squares?

Where's the Math?

In this problem, students will explore figures embedded within figures. They will explore properties of squares. As students explore the model and verbally make conjectures, they will use and develop their spatial reasoning skills and strengthen their ability to communicate mathematically. Generating the number of squares of a given size will allow children to explore and expand their number sense. Finding a pattern will allow the children to conjecture about what would happen if the pattern were extended.

SAMPLES OF HOW THE PROBLEM WAS SOLVED

The "How Many Squares?" problem was used by several teachers at Lincoln Elementary school in Macomb, Illinois. The elementary school is trying to integrate writing into their mathematics curriculum, in conjunction with the Illinois Learning Goals. Five second- and third-grade classes solved the problem in a variety of ways.

How Different Teachers Posed the Problem

Some teachers gave the problem orally with a drawing on the board. Others gave a diagram of the "stairs" formed by the squares and asked the following questions.

- How many squares can you count?
- What if we add another row at the bottom? How many squares could you count?
- Can you find a pattern that describes how the total number of stairs is growing?
- Can you tell me how to do the problem?

Among other possible questions that would generate written responses, some follow.

- How did you figure out how to count the squares?
- Do you think you have found all of the squares?
- How would explain to someone else how to count the squares?

The Setting of the Problem

Some teachers allowed children to choose whether they would work independently or in groups. Some teachers provided unifix cubes and graph paper for sketching. One teacher gave children the opportunity to work on the problem over the course of three days, leaving it and coming back to it.

Liz Bushman at Lincoln School in Dixon, Illinois did this problem as a follow-up to one where her third graders were counting equilateral triangles within a large triangle. After the experience with the triangles within triangles, Ms. Bushman's students were able to see squares nested within the figures that were larger than the smallest. One teacher gave the children colored pencils to help them keep track of the different squares they counted. She also helped them keep track of their counting using charts.

Solving the Problem

For the most part, children were able to count the 1×1 squares rather quickly. Several noticed that there were two more squares added for each successive row. Some also noticed that the number on each successive row was the next odd number after the one on the previous row. Some of the children who went on to see the larger squares that were composites of the smaller ones forgot to count the smallest. They were viewed by the teachers as being successful because they grasped the problem situation even if they missed a small detail.

There were two areas on which students tended to focus. Some focused on the pattern of adding on rows. They were able to find patterns such as adding two squares to each previous row, or subtracting two as you go up. Some talked about the fact that only odd numbers would represent rows, since they were always adding twos. This was a good aspect of the problem to focus on. One child, Ken, described how the sum of 1×1 squares went from being odd to being even, and back, in a continuing pattern (see fig. 11.1).

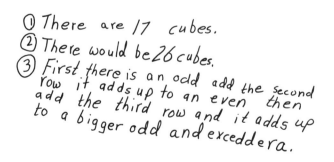

Figure 11.1

The other way that children explored the problem was to count how many larger squares were in a 4-level set of stairs. They were able to see the 2×2 squares and counted them by circling the center of each of the 2×2s that was formed by the vertices where the smaller squares came together. This was also a good aspect of the problem to focus on. Few of the children saw the large 3×3 square. It may have been that they needed a larger number of rows in order for the 3×3s to stand out (see fig. 11.2 and 11.3).

For those who counted larger squares, the coloring method that children used seemed to be very helpful in the counting process. It was very similar to the strategy that children used in the "How Many Rectangles?" problem. The colors helped children keep track of different squares that overlapped.

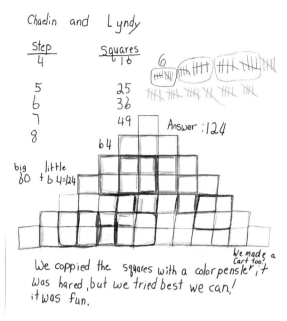

Figure 11.2

Teachers' Comments

Teachers described successful interactions with the problem on the part of their students. One teacher said, "Even though no one got the exact answer, (all missed the one 3 × 3 square), I considered the task a success. Most had learned to look beyond the obvious, and some even developed counting strategies such as circling the centers!"

Another teacher commented that it was obvious that students in her class had never seen this type of problem before. Her students all counted single squares. They also saw the pattern of how the next row was added. They were successful in exploring growth from that perspective.

Sue Holzwarth of Lincoln School in Macomb, Illinois, was the teacher who worked with her students on this problem over the course of several days. Her students are learning that a problem does not have to be solved in minutes, or even one math lesson. On the third day, after all of the children had explored the 1 × 1 squares and what happens when rows are added on, Ms. Holzwarth gave the problem back and said, "Another person and I had a difference of opinion about the number of squares that were in this picture. Someone else told me that the number we came up with was not the correct answer. Are you sure we have the correct number of 16 squares for four rows?" This question is a wonderful way to move on to day four of this problem, asking children to justify their own results.

By exploring the use of setting up charts, one teacher was receiving work from children that looked the following way. The number of squares listed reflects the sum of all of the 1 × 1 squares counted to that point.

Steps	Squares
3	9
4	16
5	25
6	36

This is a great way for children to begin exploring the relationship between the numbers in the Steps column and the numbers in the Squares problem. It is a logical area of focus for children in late elementary grades from that perspective. There are nice rules for the number of 1 × 1s, 2 × 2s, 3 × 3s, etc. if the steps have n rows. There is also a nice pattern to what the rules are.

Thanks to all the teachers who posed this problem with their children. Thanks again to Minho Kim and Mangoo Park for sharing this problem with us last year.

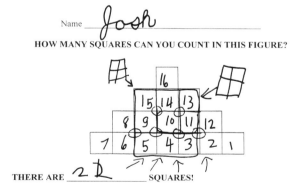

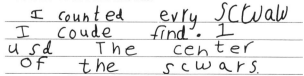

Figure 11.3

12
HOW MANY TIMES CAN YOU TAKE 1/2?

The Problem

To encourage people to attend a special concert, radio station MATH decides to give out 1242 free tickets. The station decides to distribute tickets in the following manner. Each day the station will give away half of the tickets in its possession. How many days will it take before MATH has only one ticket to give?

Range of Possible Teacher Questions
- Estimate when you think all but one ticket will be given out.
- Can you think of any ways to organize the data in order to help you think about the problem?
- Do you see a pattern?

Where's the Math?

The math in this problem situation takes several forms. Students are engaged in problem solving and communicating mathematically throughout the exploration as they discuss the problem and figure out strategies for solving it. They are also creating models as they come up with ways to show how the number of tickets is being halved. As they create tables to describe what is happening to the tickets, they are representing and analyzing data. Students are also exploring exponents in a concrete, real-life situation.

SAMPLES OF HOW THE PROBLEM WAS SOLVED

THE responses indicate that this was a good problem for fifth- and seventh-grade students. Interestingly, the responses from both grade levels included four common threads. First, the students needed time to think about what the problem was asking and needed to work to understand the first step in the problem-solving process—understand the problem. Second, the students needed to determine how to deal with a remainder when dividing by 2 and to develop a strategy, the second step in the problem-solving process, for handling remainders. Some students thought that when they obtained a remainder, they could no longer give half and, hence, the problem was solved. Others realized that a decision had to be made—either give away the remainder, keep the remainder, or alternate—depending on the situation and the numbers involved. Third, once a strategy was chosen, a solution based on that strategy was found and the process was recorded. In so doing, students employed the third step in the problem-solving process. Fourth, after choosing a strategy and obtaining a solution, students wrote about what they did. This step appeared to challenge students at each grade level, since they found it difficult to use the fourth step in the problem-solving process—look back and reflect. These observations remind us that students need assistance and experience with the problem-solving process. We also found that one problem can be used successfully at different grade levels.

Dan Brown, a fifth-grade teacher in Florida, New York, noted two common misconceptions. One was that "students thought the problem was answered when they came upon the first remainder" or "that the answer would be two days because 1242 can be given out half on day 1 and half on day 2." After dealing with the misconceptions and having a student suggest that a chart might be useful, "the students quickly became excited and quickly constructed charts to organize their information." Brown reported that his students found it difficult to explain their solutions in writing. Most students did obtain 10 or 11 days, depending on their use of the remainder and their recording process. Some students used 1242 as day 1, which resulted in 11 days. One particularly clear explanation is shown in figure 12.1. This student explained, "To get the answer of ten days I made a chart. I took the number 1,242 and divided it by 2. I didn't want to deal with remainders, so I gave them

Sara

1242

Gave Away	Day	Tickets Left
621	1	1242 - 621 = 621
311	2	621 - 311 = 310
155	3	310 - 155 = 155
78	4	155 - 78 = 77
39	5	77 - 39 = 38
19	6	38 - 19 = 19
10	7	19 - 10 = 9
5	8	9 - 5 = 4
2	9	4 - 2 = 2
1	10	2 - 1 = 1

Figure 12.1

away. When I finally got to the last ticket it was the tenth day, so I knew that had to be my answer."

The following week Brown posed another problem that would again require the students to use a table. After students clarified the problem, they applied what they had learned from the one-half problem.

The fifth-grade students in Rhonda Kinnish's class in Midland, Michigan, responded similarly to Brown's students. They employed similar strategies, drew fewer tables, but provided more elaborate written explanations. They also struggled with the remainder. One student, in an effort to make the problem manageable, reported, "First I took a calculator and divided 1,242 by 2. Then I divided the answer [621] by 2. Here I ran into a problem. My answer was 310.5. I decided to round it to 310. I did this because they can't change the amount of tickets and that is the only other option. I rounded it to 310 because it can be divided by 2. I then divided this by 2, getting 155. I then divided that by 2 and ran into the same problem, 77.5. This time I rounded up, because 78 can be divided by 2. Again I divided by 2, getting 39. When I divided this by 2 I got 19.5. I rounded again to 20 for, again, it can be divided by 2. Now I only have 10 tickets left. Dividing by 2, I got 5. Divided that by 2 and [rounded to] get 2. Divide[d] this by 2 [to] get 1. We have done this ten times therefore they had 1 left on the 10th day." Clearly, this student understands the problem and realized the complexity of dealing with the remainder. Also, the student was looking ahead when deciding how to round. Knowing that halving an even number would be best, rounding up or down was determined by which would give an even number. As the issue of whether to round up or down in this situation is arbitrary, this solution is quite reasonable.

When Ione Hesch, a seventh-grade teacher at Rockridge Junior High School (Taylor Ridge, Illinois) posed this problem, she asked the students to first predict the number of days that it would take. The estimates ranged from 8 to 59 days. That only 10 days are needed may surprise people, and it reminds us that if we had started with 2484 tickets, or twice the original number of tickets, only one more day would have been needed.

As might be expected, those who understood the problem and provided correct solutions used strategies similar to those of the fifth-grade students, but their work was more elaborate and their explanations more detailed. Many made the same errors of either listing 1242 tickets as the first day or specifying the day with two tickets left as being the last day. These minor problems with recording did not detract from how well the students demonstrated their understanding. Some created clever schemes to deal with the remainder. One student saved all the remainders "until the end" and then added the remainders all at once. This plan resulted in having tickets to give away until day 13, when only one ticket was left. Another student used a calculator and recorded the decimal value obtained by successive divisions by 2. A visual approach was used by Ethan (fig. 12.2). Ethan obtained the solution of ten days, but he made several errors in getting that answer. However, the graphical representation that he used gives a visual perspective to the process and, if done correctly, demonstrates how quickly one gets to one ticket by repeatedly halving the amount. The graphical representation can also show why starting with 2484 tickets adds only one day to the process.

Some of Hesch's comments regarding the difficulty that her students first encountered were similar to those shared by the fifth-grade teachers: "A few students did not know how to find half of a number and had to be told that dividing by 2 was one way of finding half of a number. Some students did not at first understand that there would be a smaller number of tickets to give away each day.... Putting mathematical thoughts into writing was new and difficult for many." Although these comments perhaps reflect "immature" reasoning at the seventh-grade level,

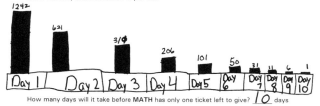
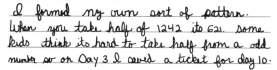

Figure 12.2

this type of investigation allows the teacher to note the level of reasoning that students possess.

What Are the Students Telling Us?

From these explanations and comments, we observe that students—

- can use the same task successfully at several grade levels.
- need to experience problems in which they have to make decisions.
- need work with strategies.
- need experiences with number relationships, such as that dividing by 2 is the same as finding one-half.
- need prodding to get beyond the initial feature— "If I get a remainder, I am done"—but can make intelligent decisions—"Round up sometimes, and round down other times."

Teachers can learn from the work of their students. Although it may be hard for students to put their thoughts on paper, teachers should make the effort to see that students have these experiences.

13
GUESS THE WEIGHT

The Problem

One event at a carnival asked contestants to guess the weight of three objects: a tetrahedron, a sphere, and a cube. Contestants are given three clues, and different combinations of the items are placed on scales. Can contestants determine the weights of the items from the information shown here?

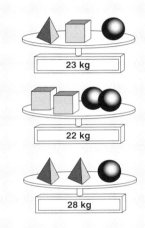

Range of Possible Teacher Questions
- What are the clues in the picture?
- What relationships can you see by studying the diagram?
- Are there any strategies you can use to help you determine relationships?

Where's the Math?

The math in this problem is strongly in the vein of algebra: students have some information and from that are asked to determine an unknown. Students are engaged in problem solving and communicating mathematically throughout the exploration as they discuss the problem and figure out strategies for solving it. They are also creating models as they come up with ways to show how the to figure out the weights. Some students may use symbolic representations to explain the relationships. Students are also exploring algebra in a concrete, real-life situation.

SAMPLES OF HOW THE PROBLEM WAS SOLVED

THIS problem generated a wide range of responses from grade 3 to post high school. Two interesting threads ran through the responses. Students, with a little help for the younger ones, understood what the problem was asking. Guess and check, or trial and error, was frequently the strategy of choice. Getting students to move beyond guess and check takes patience and a deliberate attempt on the part of the teacher. In the description of student work that follows, we see that students at the fourth-grade level, with appropriate guidance from a teacher, can and do make solution attempts using logical reasoning instead of guessing and checking.

This problem provides an ideal setting to expand student reasoning. In addition, the reasoning strategies, when written down symbolically, lead to a more formal way of solving the problem. However, as evidence from the post–high school contributor indicates, even those who have had algebraic experience can become mired in symbolism if they cannot reason.

Similar to the materials submitted by third-grade teachers Liz Bushman and Molly Kleine, all the fourth-grade students in Sandy Taylor's class used the strategy of guess and check. In contrast with most of the third-grade students who treated each diagram as a separate problem, almost all of Taylor's students understood that the value of the individual shapes would have to be the same in each diagram. Taylor established four indicators of good performance—(1) all work shown, (2) strategy used is obvious in their work, (3) connections made between equations, and (4) correct answer. However, item 3, connections made between equations, was missing in all the students' work. That is, the students did not use any information from one diagram to help them think about the relationship of the shapes in another diagram.

The fourth-grade students in Carol McBride's classroom enthusiastically worked on this and similar problems for two days. McBride wrote, "I began the first session by telling the students to put away their pencils. We read the problem and studied the clues. The blank looks on their faces were priceless! However, when we began to find what we *did* know from the clues, the blank looks began to be exchanged for sparkling interest." After discussing this problem, the students requested similar ones to solve. In addition to her problems, McBride asked her students to develop their own problems to swap and solve. She indicated that "their oral presentations of their strategies were much better than what they wrote," which is quite common for this age. Yet the two written explanations that follow indicate the depth of logical reasoning that can be exhibited by fourth-grade students.

One student wrote, "I took away half of number two so it would be eleven. I said 'eleven plus what equals twenty three?' because there is a sphere and a cube [on scale 1], so the tetrahedron is twelve. Then I went to number three and said 12 + 12 = 24, so the sphere is four. Next, I said four plus what equals eleven, so a cube is seven."

A pair of students wrote, "Every time that more tetrahedrons are on the scale, it weighs more, so we think the tetrahedrons weigh more then any other shape.

"We think the cube and the sphere weigh the same [as another sphere and cube], so we looked at the second scale and divided it by 2. A cube and a sphere weigh 11 kg; 22 ÷ 2 = 11.

"On the first scale there is a cube and a sphere, that equals eleven; 23 – 11 = 12. So the tetrahedron [equals] 12.

"Then we added 12 + 12 = 24. Then subtracted 24 from 28 = 4. So the sphere equals 4, the cube equals 7, and the tetrahedron equals 12."

When Minda Rodriguez presented this problem to her fifth-grade students, she "...wanted them to (1) draw some sort of diagram, picture, math sentence, etc., to show their work and (2) write down how they approached and solved the problem." When the work was shared with the principal, he "... asked if he could come and sit in on the follow-up discussion. To the students, this [appearance by the principal] meant that this must be really important stuff.... One student shared that she had started with the middle equation, which was met with surprise by some of the students, who had never considered starting anywhere else but the top. Mr. Shelly [the principal] shared how he himself had approached solving the problem." Two students demonstrated that they were trying to make relationships among the three scales. One student wrote, "Yes, I think the contestants can determine the weight of the items because the tetrahedron weighs the most, because if you look at the first and the second scale, the second scale has to have double of the cubes and spheres."

In Laurie Galluzzi's sixth-grade class, "Most of the students tried to get a sense of which object would be the heaviest/middle/lightest by comparing the three pictures, then they began substituting numbers." One student tried an algebraic approach, and another made an interesting table that would be an excellent way to work with the problem on a spreadsheet. The student completed the following three tables, the last one showing the correct solution.

Sphere	Triangle	Square
2	13	8
		16

Square	Triangle	Sphere
5	10	8
10		16
	20	8

Triangle	Sphere	Square
12	4	7
	8	14
24	4	

Andrea McAndrews used this task as a homework problem for her sixth-grade prealgebra class. She indicated that several students used a visual means similar to that described by McBride's fourth-grade pair of students. One student "… doubled the top picture to find the value of 'two of each shape,' then subtracted the middle picture from that to find the weight of two tetrahedrons, dividing that in half to find the weight of one tetrahedron. He then substituted into the bottom picture to find the weight of the sphere, and those two weights into the top picture to determine the weight of the cube." Although he was unable to see what to do with the information, the work of one student indicated that he combined pictures 2 and 3 to find that 2 cubes + 2 tetrahedrons + 3 spheres = 50. This procedure can be useful if one doubles the first diagram to find that 2 cubes + 2 tetrahedrons + 2 spheres = 46. From these two pieces of information, we can conclude that the extra sphere must weigh 4 kilograms, or 50 – 46.

Paul Herring presented this problem over Thanksgiving to ten members of his extended family, ranging in age from ten to the mid forties. Multiple approaches were used to solve the problem. Two of three people ages ten to twelve found the solution in about fifteen minutes. One of two people currently enrolled in calculus solved it in three minutes; the other solved it after producing two pages of figures. Two adults in their forties found the solution by using variables and equations. One person who used the same variables-and-equations logic arrived at a correct solution without writing anything down. Two adult college graduates set up equations but resorted to trial and error to find the solution.

What Are the Students Telling Us?

Students are telling us that they can handle the problem posed. They need assistance in clarifying the problem and moving beyond guess and check to think about the relationships implied within the arrangements. Teachers can use these types of experiences to help children develop appropriate symbolism that can be used to record the problem statement as well as aid in the reasoning process.

It was interesting that not all students saw a connection between the diagrams. When they did see the connection, students were intrigued by the notion of solving the puzzle.

14
WHAT ARE THE CLUES? PATTERNS IN MORE THAN ONE DIRECTION

The Problem

Can you decide what numbers should go in each of the boxes that does not contain a number? Is there a best choice for the number that would be placed in the box?

10	?	14	?
7	9	11	?
4	?	8	10
1	3	5	7

Another choice for this problem is the one below.

						*
?	17	?	?	26		
10	?	16	?	?		
?	9	12	?	?		
2	?	?	11	14		*

It doesn't become necessary for students to worry about satisfying the pattern in two directions until they are missing enough information that they need to use both rules in order to place a missing number in the box.

Range of Possible Teacher Questions
- What clues helped you select a number?
- How do you know that you've chosen a number that works here?
- Are there any other numbers that you could put here?
- Could this number show up anywhere else in the pattern?
- Is it possible to add rows or columns to this pattern?
- What other patterns do you see?
- Can you create a pattern that goes in more than one direction for the other students to solve?
- Can you project what numbers should be put in place of the *s ?

Where's the Math?

When students are engaged in exploring patterns and extending them, they are exploring algebra and number concepts. When there is a pattern in one direction, the students are exploring finding one rule. When the pattern extends to two directions, there's a rule for each direction. There is also a rule for moving along a diagonal of the two-directional pattern. That rule links the two original rules and can be described in a variety of ways. There are also other patterns that can be found which depend on the numbers in the pattern. The mathematical communication is very rich as students discuss what patterns they found and how they would describe them with a rule.

SAMPLES OF HOW THE PROBLEM WAS SOLVED

THERE was a wonderful response to this problem by members of a year-long workshop, called Project PAiRS. The teachers in the workshop have been involving their students from grades pre-K–8 in a variety of problem solving activities. They couple those with discussions and writing experiences that would allow students to discuss their findings, their strategies, and their understanding of the problem. A pre-K teacher, Sherial McKinney did the problem with her preschool group. Sherial gave the children a set of 1-inch colored tiles along with half of a page of 1-inch graph paper. The children, who she worked with in pairs, were asked to create a tiling pattern that would go across the page as well as up the page. She also explored with them the patterns that were generated along the diagonal.

Sherial found that some of the children began well and then lost interest before they filled the page. Others filled the page and began spreading their pattern across the floor. Most children were able to extend their pattern in more than one direction. One little boy knew his pattern had an error in it but had trouble determining where it was or how to fix it. Sherial made a change in one place then the child was able to go through and create consistency.

A few children had trouble seeing the pattern formed along the diagonal. Sherial would remove the tiles above the diagonal to form steps. This was very helpful in allowing children to see and describe the pattern on the steps.

The patterns that were made by most children were AB color patterns. The pattern yielded a checkerboard effect. Two of the children made AABB color patterns. Their result was a checkerboard with rectangles instead of squares. Some of the children created an AB pattern with color then placed the same color above the tiles on the first row, to give the vertical pattern a striped effect. There was some question about whether the children saw this as a pattern. It would be worthwhile to question them further about that. From her interaction with them, Sherial believed that the children were focused only on the horizontal pattern and didn't follow the idea of working in more than one direction.

When Shelley Bennett introduced the patterns in more than one direction to her fourth graders, she started with shapes. "As a warm-up activity for three days, I gave the children patterns using shapes on worksheets. I was surprised at how easily they determined the patterns."

Shelley then introduced the number patterns. She was moved by the high level of enthusiasm the children had for the number patterns. Her class discussed the patterns they discovered, clues they used to help them, and how the number patterns could be extended. Shelley shared some of the interesting points of the session.

"One of my lowest students, who has a very difficult time staying on any math task, was my most enthusiastic. Not only did he pick out the patterns quickly, he was asked by one of my highest-ability students for help."

"I observed one student filling in the top row of the graph paper before completing any of the squares near the given numbers. To my surprise it was correct. Another student filled in one row diagonally, then started at the bottom of the paper and counted up to complete the columns."

A few other teachers commented that they were surprised at which students grasped the concept of seeing patterns in more than one direction. They were the students which were described by the teachers as the weaker math students.

Donna Clausen gave the number patterns to her 7th grade class, as they were presented in the Nov. '98 issue of *TCM*. She posed the following questions to her students to write about.

1. What clues helped you select the numbers?
2. Try to find the numbers that go into the starred boxes without filling in the boxes in between.
3. Justify your starred box answers. How do you know you are correct?
4. Extend the puzzle. Where would -1 be? (How do you know?)
5. Describe as many patterns as you can see. For each pattern, why do you think it exists? Explain.

Donna's Comments:

"Students did surprisingly well. They found starred boxes fairly easily. The top square was a bit harder, but many found that all you had to do was add 10, because you moved two right and up one, so add 6 + 4 or 10. They easily located -1 by extending the pattern. They also expressed the idea that the pattern was an infinite one.

"Many had difficulty explaining why the patterns they found existed. They just said, 'That's how it works.'"

About 20% of the class was able to explain why the patterns they found existed. Donna said, "I could see many levels of understanding in the students' answers."

Patterns Included:

Add three going left to right. Subtract three going right to left.

Add four going up. Subtract four going down.

Add 7 going diagonally up from left to right. This is because you are sliding one right and one up, so 3 + 4 = 7. Subtract 7 going diagonally down from right to left. This is because you are sliding one down, and one left, so -3 + -4 = -7.

Add 1 going diagonally up from right to left. This is because you are sliding one left and one up, so -3 + 4 = 1. Subtract 1 going diagonally down from left to right. Sliding down one, and right one, -4 + 3 = -1.

If you slide right 4 and three down, you have the same number in the box. This is because 4 × 3 for each right move equals 12. 3 × -4 for each move down equals -12. 12 + -12 = 0, so you end up with the same number.

Donna's Final Reflection

"I was delighted with the results of this problem. It showed an understanding of many things including patterns, the effect of translations on the numbers, negative numbers, infinity, and much more. I would like to have them make up some problems like this to trade. I think they would like doing that. They try very hard to stump each other, so I know they would come up with some great puzzles.'

Donn Murret of Pine Elementary School in Carlsbad, California asked his two 6th grade classes to explore the problem. One group was given the pattern on paper with specific directions to fill in and extend all around the pattern. The other group was asked to copy the pattern and was given oral directions.

The students found many of the patterns. They were surprised to find negative numbers show up in the patterns. One student, Marisa described the patterns in the following way.

"I put what I put in the box because each way had a formula like when you go vertical the formula is $n + 4$. The rest of the formulas are Horizontal = $n + 3$, diagonal to the right = $n + 7$, diagonal to the left = $n + 1$. No other numbers would work in the places because each way has a formula."

Then students were asked to create their own pattern to challenge another student with. He had them create an 'answer' pattern to go along with it.

Some of the examples follow.

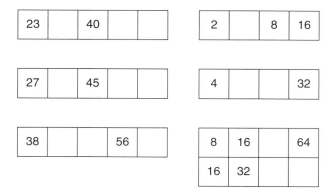

It was encouraging to see a child include a pattern that showed exponential change.

Several teachers talked about the connections that the patterns helped children make to the concepts of algebra that they were exploring.

What Are the Students Telling Us?

As is usually the case, children surprised their teachers and the editors, with their creativity and flexibility in solving the problem. Their innovative approaches told teachers a great deal about the innumerable ideas that children bring to a problem solving situation. For some of the teachers, this was the first time they presented such a problem in their class. They were encouraged by the results and especially the strengths that were shown by students who were usually less successful in their class.

That result alone allows teachers to better achieve the goals of helping children communicate mathematically, of helping children develop confidence in their ability to do mathematics, and of helping children become problem solvers. It also reinforces the fact that all children can do mathematics.

15 WRIST BANDS

The Problem

Students from two classrooms at Euclid Elementary School were going on a trip to the Mathland Museum for Honors Day. Seventeen of Mr. Alpha's students and fourteen of Ms. Gamma's students were chosen to attend. The students wanted to wear something with their school's name on it. The principal, Ms. L. C. Multiple, agreed to give the students 150 wrist bands left over from the school carnival if they could solve a challenging problem. Each wristband was white with five numbered circles. If each student wears a wristband and different numbers of circles on each band are colored with only the school's color—red—can each band be colored so that it is different from every other band?

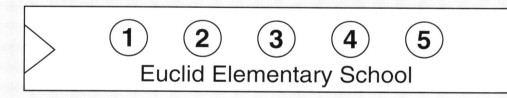

Range of Possible Teacher Questions

- What would two different wristbands look like?
- How can you organize the data to keep track of the different types of wristbands?
- Can you see a pattern?
- Can you explain the role of color in the wristbands?

Where's the Math?

This problem has many mathematical facets. It is a counting problem similar to "Decoration Delight" (exercise #7, on page 24 of this book). Students can find interesting patterns in their counting process, depending on their listing strategies. Discrete mathematics is a branch of mathematics that includes such topics as counting problems of this type.

SAMPLES OF HOW THE PROBLEM WAS SOLVED

The student work submitted for "Wrist Bands" demonstrates that children understood the problem and that several recognized the need to organize their thoughts so that they could find all thirty-two possibilities.

Kristina Kalb submitted work from several fourth-grade students. Kelsey worked the problem "by separating the combinations into groups" (see fig. 15.1). As might be expected, Kelsey did not list the wristband that had zero circles colored. Note that, including the missing band, Kelsey generated the string of values 1, 5, 10, 10, 5, 1 for zero, one, two, three, four, and five circles colored. This string is the sixth row in Pascal's triangle.

Several other students also demonstrated excellent thinking and found all the possible ways that the wristbands could be colored. Andy found all the possibilities and stated, "I know that this is all of the wristbands because there are no more combinations that you can make, unless you divide the circles in half or in thirds, etcetera. Also, if you had six circles, you could make more."

Allison demonstrated insight, as well. She wrote, "I brainstormed patterns, and then I did the exact opposite of colors. For example: RRRRR, then you can do WWWWW." She clearly saw a powerful connection between pairs of colored bands and used this connection to move from her original thought of thirty-one wristbands to all thirty-two.

Brenda Frankel offered an excellent description of the work and thinking that went on in her fifth-grade classroom when this problem was posed:

> Many students were immediately disturbed by the fact that *150* wristbands were available for use. "Are we supposed to find 31 different wrist bands or 150?" they queried. I, myself, was unsure if the number 150 was provided as an exercise in the elimination of unnecessary data or for some other purpose. I reassured them that they were seeking 31 different combinations of colored circles.

Once the students began working, Frankel was able to discern several approaches that students used as they tried to solve the problem. She indicated that some students worked the problem by listing combinations without a systematic approach or wrestled with the structure of the problem.

> Some students set about their task randomly trying to find 31 different combinations with no apparent system. Friedi and Malka started out by arranging the *numbered* circles into different patterns rather than using different *colored* circles. When I clarified the problem posed, they left the numbers as they were but went back and successfully found 31 different patterns with no apparent system.

By far, most students realized that a strategy was needed to accomplish this task. Some interesting questions arose: Avigayil attempted to increase the potential patterns and asked, "Am I allowed to color a fraction of a circle?" One student having difficulty checked whether she could "color outside the circles," and numerous students questioned whether a blank pattern could be included in the total.

Frankel also gave several examples of student work that made use of an organized approach to solving the problem. These examples demonstrate the sophisticated approaches and insights that fifth-grade students can employ.

> Miriam and Ester neatly and systematically depicted their pattern. They began with variations of one colored circle and proceeded to all five circles. Pamela and Shoshana numbered each wristband and were able to efficiently refer to each wristband as they described what they had done. They began with five colored circles. They continued to variations of four, three, one, and then two. They were stuck for bands #30 and #31. They found a solution for #30 and used a noncolored band for #31. They [believed] that #31 did not conform to the rules and stated rather sadly that "we could not solve the problem."

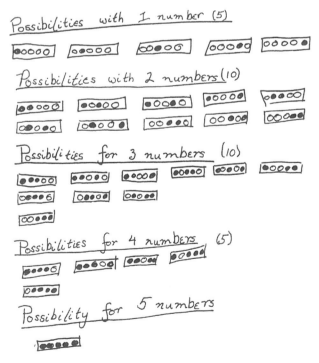

Figure 15.1. Kelsey's grouping of wristband colorings

Nava and Katherine ... made statements about their insights, "[four colored circles] are simple because there are not many opportunities, and [five colored circles]: obviously, only 1. To shorten this you could skip 5 and 1 because they are obvious."

Frankel concluded with a comment about her students' accomplishment and enjoyment of the task: "My class enjoyed this exercise tremendously. Most of the students completed the task. There was an excitement and energy throughout the room."

What Are the Students Telling Us?

Students are telling us that they can handle the problem posed. They need assistance in clarifying the problem and moving beyond guess and check to think about the relationships implied within the arrangements. Teachers can use these types of experiences to help children develop appropriate symbolism that can be used to record the problem statement as well as aid in the reasoning process.

It was interesting that not all students saw a connection between the diagrams. When they did see the connection, students were intrigued by the notion of solving the puzzle.

16
TAKE TWO: FAIR OR UNFAIR?

The Problem

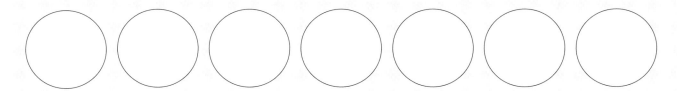

Play the strategy game "Take Two": Place seven chips in a row. Two players take turns removing one or two chips each turn. The person to remove the last chip is the winner.

Is "Take Two" a fair game? In a fair game, each player has an equal chance of winning. This game can be played with chips, two-colored counters, checkers, cubes, buttons, pennies, squares of colored paper, etc. Although the students may need help understanding the problem and the constraints under which they are to work, avoid giving too much guidance. View this task as more than an exercise for which students are seeking a correct answer. Ask your students to make and test conjectures. Ask them to record their conjectures as they play the game more times.

A nice whole-group exploration would be to collect the results of games played by pairs around the room and to see if there are trends. A variety of ways of representing the data will help the class determine whether they think the game is fair. In an alternative version of the game, the winner forces the opponent to remove the last chip.

Range of Possible Teacher Questions
- Does it make a difference who plays first?
- At what point in the game can you tell who will win the game?
- Is there one of you who is winning more than the other?
- What can that person share about his or her strategy in playing?
- How would the game change if players used eight or nine chips?

Where's the Math?
Game theory is a separate branch of mathematics and economics. The question of whether a game is fair is often studied in terms of probability; that is, does each player have an equal probability of winning in a given round of the game? "Take Two" is a variant of the game nim. Playing the game allows children to explore concepts of addition, subtraction, logical reasoning, problem solving, and spatial visualization. Children have opportunities to do multistep problem solving as they decide what the possible outcomes of a move would be. The game is mathematically worthwhile even if your students do not reach the point of answering whether the game is fair.

SAMPLES OF HOW THE PROBLEM WAS SOLVED

Barbara Lemme and her twenty-one second graders played the game several times. Their goals were to attempt to devise strategies for winning as well as to determine whether "Take Two" is a fair game. Lemme reported, "My students ... readily involved themselves in the game and played it many times over with their partner." Lemme also said that the children formed some interesting conjectures to explore as they played the game.

One issue that arose from the explorations was that the children held different views about what it means for a game to be fair. Several children described the game as fair because they had not cheated or they had waited their turns. These interpretations resulted in a discussion in which the class compared the concepts of *fair game* and *fair play*. "I was surprised at how confusing the terminology of games can be," commented Lemme. This observation is a good reminder that teachers might have different perspectives on classroom issues than children do. Through discussion, we can assess more accurately how children interpret what is being said.

The writing of two students, Paul and Nate, shows how these children viewed the concept of *fair game* from the perspective of *fair play*:

- Paul wrote, "Yes the game is fair because if you go first and take one and the other player takes two, then you win that game. I let the other player go first, and he took two and I took one and won the game."
- Nate wrote, "I think the game is fair because I do not cheat."

Two other children gave a different reason that the game is fair:

- Sakara wrote, "It was fair because both people had a chance of winning."
- Alexandra wrote, "I think the game is fair because we have the same chance of winning."

Both Sakara and Alexandra won much more often than the children against whom they were playing. Sakara continued, "I found a better chance of winning. I noticed when I went first, Elizabeth kept winning. I thought that if Elizabeth went first, I could win, and I did." Alexandra made the same comment about being the player to go second and winning as a result. Paul found that he could win if he went second, as well.

Jacob said, "This game is not fair. James kept winning. Then I saw what he was doing. Then I tried it. Then I won. The first player takes two away. Then you take one away. Then there are three. He takes one or two. You win." Jacob is actually off by one token; however, if each player takes one after the first move, the pattern of removals would be two-one-one. Then three tokens are left, and indeed the first player would win.

One child wrote, "My partner kept on winning until I said, 'How about I go first now?' He said OK. So I went first and kept on winning." This child did not indicate whether the game is fair, but the idea that the player to go first would win was definitely clear.

Several children were able to win if they went second, thus clouding the issue of whether going first is an advantage. Further exploration would be required for some of the children to determine whether a particular path would lead to a win. An interesting follow-up step would be for the children to begin to collect data about what combination of moves leads to a win.

Nate described his strategy for winning as follows: "I take two when I go first, and I make it so there are three blocks left so I win." This strategy assumes that the person who goes second will take one token rather than two. The first player can guarantee a win by removing one token on the first move. Table 16.1 shows the possibilities. The goal is to do as Nate suggests: "I make it so there are three blocks left so I win." At that point, no matter what player 2 does, player 1 will take the last token.

In playing this game with kindergartners and first graders, I observed three phases in the development of their problem-solving skills. In the first phase, children see little connection between how many tokens they take and the outcome of the game. These children tend not to look beyond the current move. In the next phase, children begin to look ahead one move and see how one move may affect the next. In the third phase, children begin looking at how the first move will affect the outcome of the game. Looking ahead and considering the outcomes of what happens first is important in multistep problem solving.

In a recent experience of playing this game with a kindergartner named Nicholas, we used pennies. The first time we played the game, when we got

Table 16.1
Possibilities for Winning "Take Two"

	Move 1 Player 1	Move 2 Player 2	Move 3 Player 1	Number Left
Game 1	1	2	1	3
Game 2	1	1	2	3

down to three pennies left and it was my turn, Nicholas began laughing. I asked what he was laughing at, and he said, "I'm going to win!" I asked how he knew that, and he said, "If you take one, I get two. If you take two, I get one." After playing three times, Nicholas began counting how many pennies were left before he took his turn and would determine whether he could take enough to leave three. After playing five times, he suggested that we play "take three or less." He immediately took three pennies and grinned at me. When I asked why he was smiling, he explained that he had won.

I suggested that we play "Take Four or Less" to see whether he would take two pennies to leave five, thus ensuring that he would win. He took one. He had not yet extended his strategy to this level.

After playing "Take Four or Less" a few times, we went back to playing "Take Two." I asked Nicholas if it mattered who went first. He said that it did because he liked to go first. When I asked whether I would win if I went first, he said that I would not necessarily win. He had not recognized that the player who went first could control the game.

What Are the Students Telling Us?

"Take Two" is a good problem to explore because it has several entry and exit points. The fact that the game is not fair does not detract from its value. In fact, it gives children the opportunity to reason about a game and determine whether it is fair.

"Take Two" is one possible way to explore and talk about the likelihood that an event will occur in a situation that directly interests and involves young children. They have personal experiences with the game that can be brought to the discussion. They can gather data about the different moves and the resulting outcomes, recording what effect going first has on which player wins. Finally, they can explore problems related to how one wins the game.

Thanks to Barbara Lemme and her second-grade class at Wawaloam Elementary School in Exeter, Rhode Island, for exploring this problem and sharing their input with "Problem Solvers." Thanks also to Nicholas in Macomb, Illinois.

17

WHAT SHAPES CAN YOU MAKE?

The Problem

Use four isosceles triangles as shown, and place all of them together along edges with no overlapping. How many different figures can be made? One example is shown.

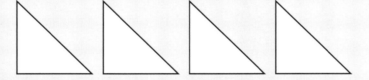

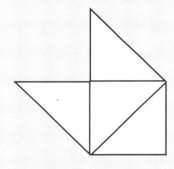

Range of Possible Teacher Questions
- What strategies are you using to find shapes?
- If you turn or flip a figure, do you get a different figure?

Where's the Math?

This spatial problem requires that students manipulate shapes. Students must consider congruence to determine whether a new arrangement is really different from others. Classification skills can be practiced as students sort their shapes as triangles, quadrilaterals, pentagons, and hexagons. Subclassifications of the quadrilaterals—rectangles, squares, parallelograms, and trapezoids—can also be considered. In addition, this problem is about counting. Students can investigate whether a pattern exists between the number of triangles used and the number of shapes that can be made.

SAMPLES OF HOW THE PROBLEM WAS SOLVED

STUDENTS from prekindergarten to high school and their teachers found the "What Shapes Can You Make?" problem to be interesting, motivating, and fun. The variety of responses that we received demonstrates the rich potential of this problem to challenge spatial sense, extend understanding of geometric shapes, enhance classification skills, encourage classroom discourse, provide opportunities for group work, and allow students to express their creativity.

Two of Sherial McKinney's prekindergarten students produced ten different shapes with some assistance in putting sides together so that they "touched side to side and not point to side and had the same length." One student put the four triangles together as a large triangle without assistance, and a second student produced a rectangle with some assistance. The students repeated some shapes and needed help to turn them to recognize that they looked like shapes that they had previously made.

Laurie MacKay, a third-grade teacher, and Karen MacKay presented this problem to five- and six-year-olds in Karen's first-grade class. They asked the children to glue their four triangle shapes onto another piece of ten-by-twelve-centimeter grid paper, then cut around the outline of the four-triangle shape so that the four triangles were still visible. As the children made new shapes, they related them to their environment and known objects. Laurie MacKay and Karen MacKay sent the examples in figure 17.1 to share some of the children's ideas for their shapes.

Although many children were confused about the criteria and used more than four triangles or arranged the triangles with only their corners touching, others worked with the criteria in mind right away. One child commented, "It's like making a puzzle." The teachers noticed that once one new combination was discovered, the children suddenly realized that many new shapes were possible. As the work progressed, one child commented, "Some of our shapes have to be the same because people have the same ideas."

The MacKays's reflections led them to wonder whether the results might have been different if the children had first spent some time using the same triangles to create *any* shape they wished. Then the "What Shapes Can You Make?" problem would have been an extension of the initial exploratory activity, with added restrictions.

Rebecca Anderson is a consultant who works with programs for gifted students in kindergarten through sixth grade. She participated in Ann Whan's and Kim Erichsen's second-grade classrooms for one week. The children began the week by using tangrams and geoboards to explore ways to make different sizes and shapes of triangles. Through these lessons, the students were introduced to the shape of a parallelogram.

Anderson wrote, "Many of the students [thought that] *parallelogram* was too difficult a word to remember. We discussed other big words that they might know, such as *brontosaurus*, *tyrannosaurus rex*, and other dinosaur names. The students agreed they could remember those types of names and were willing to try to remember the name parallelogram."

Anderson observed that Logan and Nathan wanted to discuss only shapes that they could recognize and name as a familiar shape or polygon, such as *triangle*, *square*, and *rectangle*. After making these familiar shapes with their triangles, they believed that no other shapes were possible. They continued to manipulate the triangles, when Logan called out, "Oh, oh, we've got one! Mrs. Anderson, we've got one!"

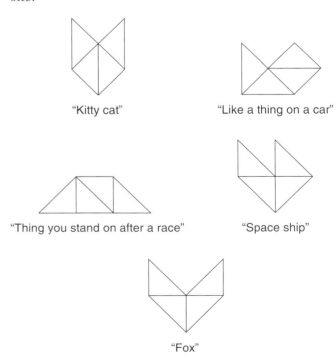

Figure 17.1. Children related their shapes to familiar objects.

"What have you got, Logan?"

"Ah, ah … brontosaurus?"

"No," Nathan replied, "it's a eucalyptus?"

In unison, they replied, "No, but it's one of those big words."

"A *parallelogram*," Mrs. Anderson replied.

"Yes!" They were thrilled by their discovery.

When they made a trapezoid, Logan said, "I don't think we can use that. It doesn't have a name." Anderson discussed with the students whether all shapes had to have specific names. The two boys agreed that it was fine not to have a name for a shape but said, "We like those [with names] better." They were relieved to find that their new shape had the name of *trapezoid*.

The students in Ann Whan's class chose to name their shapes *crown*, *bed*, *car*, *wolf*, *castle*, and so on. Two students drew details on the shapes so that the dog and cat had a tongue and whiskers. The students were divided equally in their choice of outlining the shape or of tracing each individual triangle. The discovery of the parallelogram was again a notable event. Spencer said, "Look, a parallelogram! I'm getting the hang of it. I've been making them at home at night with my triangles."

The teachers found that they were able to identify visual spatial skills in their students, especially in some of the computationally weaker students. They also noticed that this problem challenged some students who usually found learning to be easy and were not used to persevering at a task.

Angela Wilson's second-grade students explored arranging the four triangles over a three-day period. As with many students in the early grades, they focused on creating shapes that they could identify as something familiar. Each of the groups made a polygon book of their shapes. Figure 17.2a shows one group's book, and figure 17.2b pictures number 11 of their fourteen shapes.

Creativity abounded in Julie Bice's third-grade classroom as groups created and named shapes over a three-day period. The pictures of the groups' work illustrate ways that the third graders approached this activity. This problem motivated Bice and her students to use their imaginations as the shapes were created. (See fig. 17.3.)

Mary Kay Varley's third-grade students had already worked with tangrams, pentominoes, and hexiamonds, so she introduced this problem to her class from that perspective. (See "Math by the Month: Geometry" in the February 1999 issue on page 346 for a brief introduction to pentominoes.) They began by making predictions and learning some vocabulary. The vocabulary words that they

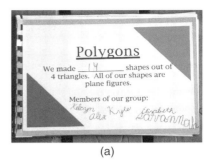

(a)

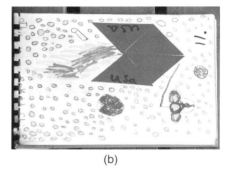

(b)

Figure 17.2. Second graders made a book showing their fourteen shapes

discussed were *congruent*, *isosceles*, *right angles*, *rotation*, *flip*, *reflection*, and *vertex*.

Possibly drawing on their knowledge of pentomines, most of the students predicted that they would find twelve shapes. A few students predicted that they would find eight or ten shapes because they

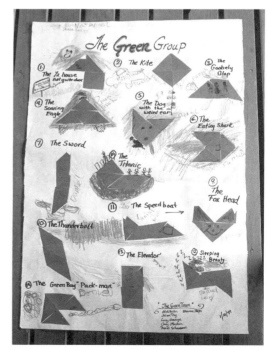

Figure 17.3. More creative interpretations of the shapes are shown.

were working with only four triangles. Another student predicted that sixteen shapes would be found because 16 is a multiple of 4.

Students found the basic geometric shapes fairly quickly. The triangle, square, and rectangle were found first, but a few students immediately turned and flipped their triangles to find the more unusual shapes. With some shapes, they became confused as to whether a shape was unique when the triangles within the shape were oriented differently. (See fig. 17.4.)

Varley wrote, "This confusion led us back to a discussion of what *congruent* means. The thin, narrow parallelogram was difficult to find because [students] found the wider one first. They did not recognize at first that the two were different, and they weren't expecting to find two shapes so similar." Many groups wanted to stop after finding twelve shapes, thinking that they had proved their prediction. When a student discovered a thirteenth shape, however, the search was on again. It took them another hour to find all fourteen distinct shapes.

Varley and three of her teammates sponsor an after-school activity known as Club M.A.T.H. The "What Shapes Can You Make?" problem was extended after school to classify the fourteen shapes in different ways, such as those having the same number of sides, those containing right angles, and those having mirror symmetry. The students noticed that many of the pieces fit together nicely and immediately began creating shapes and puzzles.

Students also used the shapes to create tessellations. This activity led to the idea to create a patchwork-quilt pattern using one of the shapes on each of four patches. In planning how to use their shape to make a four-block quilt patch, students had to work together to decide how they would rotate or flip their patch to form a symmetrical and pleasing design. (See fig. 17.5.)

After Shelley Bennett's fourth-grade students created their shapes, she asked them to classify the figures. Some groups classified their figures by colors, whereas others classified them by the names that

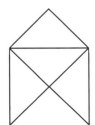 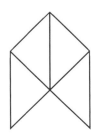

Figure 17.4. Students struggled with the notion of congruence.

Figure 17.5. A patchwork-quilt design was formed by rotations of one shape

they had given them, such as *animals* or *buildings*. Very few groups thought to classify them by the number of sides. Finally, Bennett asked them to put the shapes into four groups: triangles, quadrilaterals, pentagons, and hexagons. This request helped guide the students who had not created all fourteen shapes. Bennett was surprised to see that some of her top students in computation were laboring over finding a variety of figures. She also noticed that a student with attention deficit disorder was working harder than he had all year.

When Minda Rodriguez's students became impatient and wanted to know how many figures they could make, Rodriguez asked the groups to raise their hands if they had found more than five figures. Since all groups had identified at least five figures, she asked whether any groups had found ten figures. One group had created more than ten, so she asked whether any group had gotten twenty figures. Since no group did, Rodriguez told the students that they could expect to find more than ten but less than twenty figures.

Marjorie Adamczyk's classes of fourth and fifth graders had just explored examples of similar and congruent figures when she posed this problem. Her students used cutout paper triangles in exploring the problem. Lauren was puzzled by two drawings of squares that she had made. One was rotated 45 degrees, and when asked whether they were the same, she said, "I know they are the same shape, but one is turned." Adamczyk asked if Lauren remembered what word describes that relationship. Lauren first thought back to the word *similar*. She said

"Similar ... 'c' ... 'con' ... congruent" to arrive at the word she needed.

Mike and Alex had a conversation about the size of the triangles. Alex said that he wished he could cut the triangles in half again. Mike agreed, but seconds later, Mike told him that halving would not matter, it would produce the same shape, just smaller.

Adamczyk wrote, "The issue of congruence is still troubling several students in the afternoon class, since they drew the same figure in turned positions. Thinking about this activity later, I wondered if the results were different because of the fact that the afternoon class is more competitive and may have been competing rather than problem solving. One girl used the entire time to make as many shapes as she could with little regard to whether she had already drawn the figure."

With a seventh-grade-honors class, Susan Hallmark was able to discuss the "no overlapping" restriction as meaning that a leg of a triangle had to touch a leg of another triangle. Also, a hypotenuse had to touch another hypotenuse, but a leg should not touch a hypotenuse. Her students also decided that *different* meant that a figure could not be reflected, flipped, or rotated.

After the class made several figures at random, they divided them into four categories. They reasoned that if they had only three triangles to use, they could make only four different figures. Those four figures served as bases for the four categories. The three triangles remained stationary, and a fourth triangle was added in various places. Each time that the triangle was placed in a new location, the students made sure that they did not already have that figure. The students found twenty-one figures and were fairly certain of their answer.

Hallmark took the chart that the students had created to share in a class on "Mathematics for Gifted Students" in which she was enrolled. As the chart was examined, duplicates were identified. The seventh-grade students were pleased that other educators were interested in their work, and they created a final chart showing all the shapes for Hallmark to share. (See fig. 17.6.)

Sharing the seventh-graders' chart in Hallmark's class led to further exploration by Paul Herring, a high school teacher. Herring asked how many shapes could be made with five triangles. Through working with his eleven-year-old son and using a similar strategy, the seventh-grade chart was extended to thirty-three shapes and brought to class for scrutiny. After one duplicate was removed, it was determined that thirty-two different shapes were possible with five triangles.

Figure 17.6. Seventh graders created a chart showing all fourteen shapes.

What Are the Students Telling Us?

The level of enthusiasm, learning, and discussion about this problem has been exciting for both teachers and students. This problem can be explored from prekindergarten through high school. It includes a variety of strategies and results. Learning about geometry really is fun!

Reference

Rectanus, Cheryl. *Math by All Means: Geometry Grades 3–4.* Sausalito, Calif.: Math Solutions Publications, 1994.

18
COUNTING CUBES

The Problem

Eric loved to count. One day Ms. Fox dumped some cubes on Eric's desk. As might be expected, Eric began to count the cubes. He reported the following to Ms. Fox. "When I count the cubes by three, I have one left over; when I count by four, I have one left over; and when I count by seven, I have none left over." From this information can Ms. Fox determine how many cubes Eric has?

Range of Possible Teacher Questions

- What are the clues in the problem that help you?
- What can you tell about the number from what Eric said?
- Is your solution the only possible one?

Where's the Math?

This problem has many mathematical facets. It involves skip counting and implies division with remainders. Problems like this one can be solved by using the concept of least common multiple. Of course, the notion of skip counting by twos, threes, fours, and sevens is also involved.

SAMPLES OF HOW THE PROBLEM WAS SOLVED

THE responses to "Counting Cubes" illustrate how the same problem can be used successfully with children of different ages and at different grade levels. They also illustrate that a problem with a seemingly simple answer can be explored to a greater depth with appropriate coaching. We received several responses indicating that children think like mathematicians at all grade levels.

Beginning at first grade, children showed that they could understand a problem with several conditions and organize their work. Some students indicated that they could use sophisticated topics, such as square numbers, at an early age. One fourth-grade student showed that she could use a spreadsheet and organize the information in such a way as to generate all solutions to this interesting problem. However, we also see that the structure provided by teachers, along with their reflective analyses, are important components in helping students become mathematicians.

Chree Perkins introduced the problem by reading the book *A Remainder of One* (Pinczes 1995). Her students regularly do group problem solving and had been introduced to the ideas of multiplication, skip counting, and prime numbers. "One child thought that the problem could not be solved because we did not know how many groups of seven Eric had put them into," she explained. However, one group that had "totally mastered the concept of multiplication and square numbers and prime numbers" solved the problem. One group member immediately said, "Well, if we knew how much 72 is, we could figure it out." This insight provided direction to decide that 49 cubes could be the answer.

A first-grade teacher, Marianne DeRise, "thought that they should be able to handle this problem, since we have done similar activities.... Within one-half hour just about everyone had gotten it." But DeRise found that her students had a hard time explaining their answers.

Sue LeBeau's fifth-grade students thought that this problem was interesting and that they could solve it but were not sure whether more than one solution was possible. LeBeau noted, "Although all students who were working on the solution have been exposed to the same problem-solving strategies and skills throughout the year, every group used a different strategy." Figure 18.1 shows one strategy.

twos	threes	fours	sevens
2	3	4	7
4	6	8	14
6	9	12	21
8	12	16	28
10	15	20	35
12	18	24	42
14	21	28	(49)
16	24	32	56
18	27	36	Lauren
20	30	40	abby
22	33	44	Bryan
24	36	48	
26	39		
28	42		
30	45		
32	48		
34	51		
36			
38			
40			
42			
44			
46			
48			

Figure 18.1. One fifth-grade strategy involved using a table.

LeBeau concluded, "All groups arrived at the same answer, 49; however, the strategies that each group employed proved to be enlightening for both students and teacher. These problems certainly help not only my students in their problem-solving strategies but me, as well, to understand just how students think about problem solving."

Peggy A. O'Conner teaches thirty-two accelerated fourth-grade students and works with problem solving on a daily basis, including assigning a nonroutine homework problem each night. She explained, "Though I thought that Eric's problem was not a particular stretch for my class, I discovered a problem-solving behavior of my class that was a bit disturbing. I noticed that once students had strategized, found a solution, and explained their process, the search was over. Exploration of other solution possibilities was not considered, and in this case, other solutions had a bearing on how to answer the posed question.

"Most children understood that combining clues led them to conclude that the number of cubes needed to be a multiple of 7 and odd. The most common strategy, then, was to write the multiples of 7, cross out the even numbers, and begin eliminating other multiples through division by 3 and 4, paying attention to the remainders (fig. 18.2). All my class discovered that 49 cubes were dumped onto Eric's desk.

First we knew that the number had to be odd and a multiple of 7. It has to be odd because there has to be 1 left over when you count by 3, 2, and 4. It has to be a multiple of seven because there has to be none leftover. We first tried 7, but when we counted by 4 there was 3 leftover. When there is supposed to be 1. Then we tried 21 (We skipped all the even numbers.) but when we counted by 3 there was none leftover. So we tried 35. But when we counted by 3 there was 2 leftover. Then we tried 49. We checked if it was right with 3 and there was 1 leftover. Then we checked for 2 there was 1 leftover. We tried 4 there was one leftover. Then tried 7 and there was none leftover. And it was an odd number. So yes from that information Miss Fox can figure out how many cubes Eric has. We used a calculator to do the math.

⭐ 49 cubes *and mental math!*

Figure 18.2. Doug and Bobby typed their solution.

The search was over. To a person, the children believed that Ms. Fox could determine how many cubes Eric had. No one realized that there might have been another quantity of cubes on the desk, which would then question the certainty that Ms. Fox knew how many cubes there were."

O'Connor then drafted a letter to her students. After complimenting them on their work, she wrote, "My dilemma is this: Are you absolutely, positively, ultimately sure that Ms. Fox can determine *exactly* how many cubes she spilled onto Eric's desk? Just because you came up with a solution, does this mean that this IS the solution? Can you broaden your thinking and revisit this problem?"

Describing the search for additional solutions, O'Connor found that "[t]heir approaches were more random than refined; however, most students did find an additional amount of cubes.... Still, once finding that additional amount, the desire to stop was prominent" (fig. 18.3). Several students then considered whether some of the answers would be reasonable for the problem. As one student wrote, "No teacher in her right mind would drop 133 cubes on a student's desk."

O'Connor noted, "Only two students used sophisticated strategies to develop an organized, indisputable explanation to their conclusions." One of these solutions is found in figure 18.4.

> My solution is that Ms. Fox can't determine how many cubes Eric counted because 133 and 49 fit the clues. I listed the multiples of 7 up to 154. 133 and 49 fit the clues.
>
> Ms. Fox can not be sure exactly how many cubes she spilled onto Eric's desk because there is more than one correct answer. Beginning with 49 (the previous answer) and adding by 49, you will come to the number 637. This is also a correct answer. 3 goes into 637 212 times with one left, 4 goes into 637 159 times one remaining and 7 goes into 637 evenly 91 times.

Figure 18.3. Students extended their work.

Re: Eric's Problem

I solved the dilemma using the following method:

In the left column of a spreadsheet, I listed the multiples of seven (7, 14, 21, …).

In the next three columns I calculated the remainder of the number from the first column when divided by 2, 3 and 4. The value in each of these columns I set to one (1) if the remainder equaled one. Otherwise, I set the value to zero.

In the fifth column, I added the values from the remainder columns. I then sorted by this column. All of the numbers whose sum of the remainder columns equals three satisfy the problem.

Strangely, all are equal to 49 + a multiple of 84.

49+(84*0)=49
49+(84*1)=133
49+(8.)=217

and so on.

The answer to the dilemma is that I am absolutely, positively, ultimately sure that more information is needed to determine exactly how many blocks Ms. Fox dumped on Eric's desk. It could have been 49, 133, 217 or infinitely many more answers limited only by the strength of the desk to hold the blocks and depth of Eric's resolve to count them.

Spreadsheets are great because they allow many calculations to be done very quickly. But I am not certain I understand the mathematical concept Ms. Fox is looking for.

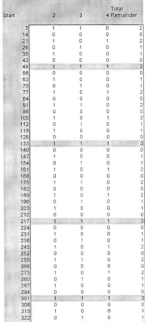

Figure 18.4. Megan used a spreadsheet in her work.

What Are the Students Telling Us?

These examples demonstrate that students can "think like mathematicians." The work of the students in O'Connor's class indicates how important it is for teachers to continue to ask students to verify solutions and to make sure that they have considered whether they can find other answers to a problem. This habit does not come naturally for adults or children. However, we see that children can grapple and have success with problems involving mathematical concepts that they are just beginning to learn. Our work as teaches is to continue to expand the range of situations in which children can perform like mathematicians.

Reference

Pinczes, Elinor J. *A Remainder of One*. Palo Alto, Calif.: Houghton Mifflin Co., 1995.

19
THE ORANGE GAME PROBLEM

The Problem

Rudy had a dream that he was on a giant orange and yellow checkerboard. Each square on the checkerboard was one foot long and one foot wide. As he stepped on the first square in the corner, two oranges appeared. Rudy loved oranges! He was so excited that he picked them up and put them in his backpack. He walked to the next square. As his foot landed, four oranges appeared on that square. They were so bright and fresh, Rudy quickly put them in his backpack with the first two. As Rudy stepped on the third square, eight oranges appeared. He was in an orange game! He added the eight oranges to the ones already in his backpack. His backpack was getting fuller and heavier! As Rudy continued to walk onto squares, oranges kept appearing in the same way that they had before. He would add them to the ones in his backpack and move on. But the backpack was getting heavy and Rudy was getting tired.

Range of Possible Teacher Questions
- How many oranges did Rudy have by the time he got to the seventh square?
- Can you describe how the number of oranges was growing as Rudy stepped from any one square to the next?
- Could he carry all of the oranges he collected? Why or why not?
- When would Rudy's backpack be full if it could hold 20 oranges?
- Explain how you figured out the answers above.
- How many oranges do you think you could carry?
- Write a story about what you think would happen when Rudy stepped on the ninth square.

Where's the Math?

This problem has several mathematical themes. The first is related to doubling. The notion that the number of oranges is doubling does not at first appear challenging to children. But what teachers learn from listening is that some children have the misconception that adding two is doubling. This problem gives a concrete experience through which to discuss the concept of doubling. Also related to doubling is the notion of how fast the number of oranges begins to grow when one is doubling. In that way this problem is similar to the allowance problem. Another aspect of this problem relates to measurement of volume with a nonstandard measure—oranges. Just how much space do the oranges take up on the ninth square? Of course, there is problem solving in this problem and estimation if one determines how many oranges they can carry.

SAMPLES OF HOW THE PROBLEM WAS SOLVED BY OTHER CHILDREN

SANDY Taylor, a fourth grade teacher in Macomb, Illinois gave the "Orange Game" problem to her students. The class began the problem by discussing the story and sharing the strategies they thought they might use to solve the problem. The group eventually decided to put a checkerboard and a table on the chalkboard. At that point the children worked individually on the problem. Most students used a table to record their data and looked for a pattern.

To answer the question of how many oranges Rudy found on any given square some successful approaches used by several students were:
- Multiplying the number of oranges on the previous square by two to get the number of oranges on the next square. For example, if there had been two on the first square, the next square would have two times that, or four. The number on the third square would be found by multiplying four by two.
- Numbering the squares and writing down as many twos as the number on the square, then multiplying. For example, on square four, $2 \times 2 \times 2 \times 2 = 16$, so on the 4th square he would get 16 oranges.

Two misconceptions about how the problem should be solved showed up in several students' work. They are variations on the same theme. Several students described the problem as an "add two" problem. The two approaches are described here.
- Getting two more oranges on each square. For example, he'd start with two and add two to the previous number on each successive square. So to figure out how many were on the eighth square, he would add two more to the fourteen he already had, for a total of 16.
- Using the number of the square to generate how many oranges he had total. If he's on the eighth square, there would be eight twos added together, $2 + 2 + 2 + 2 + 2 + 2 + 2 + 2 = 16$ oranges.

In several instances, children began by doubling the previous number and during the process switched over to adding two to get the next number. Others focused on how many Rudy found on a given square, but did not explore the problem of how many oranges Rudy had total.

Several of the children who understood the problem as a doubling situation went on to determine how many oranges Rudy would have altogether, by the ninth square. They were very successful in determining that he would have 1022 oranges. A few of the children found a pattern for summing the total number of oranges Rudy had. They noticed that when adding, the sum of all of the previous squares would be two less than the number on a given square. For example, when he stepped on the 5th square, Rudy would get 32 oranges. The total number he would have would be 32 plus 30 from the previous squares, which would be 62 oranges. Some of the children used this strategy to determine that on the 9th square he would get 512, and they'd know that the sum of the previous squares would be 510 without adding the numbers up.

Very few of the children wrote a story about what happened on the ninth square. They just told how many oranges he'd have by that time. However, one child said that Rudy fainted on square nine, because he couldn't carry 1,022 oranges. Another child explained that backpacks for carrying the oranges began appearing as well. Some of the other children said he woke up when he got to square nine. The children in Ms. Taylor's class seemed very focused on solving the problem. They were not as interested in writing the story. Very few students made more than a brief comment when asked to write a story about what happened to Rudy on the ninth square.

Amy Benedict solved this problem with the students in her 6th grade mathematics class at Hamilton Middle School in Hamilton, Illinois. She posed the questions the way they were posed by the author. Ms. Benedict had students begin exploring the problem individually. After about 15 minutes of exploration, students could continue to work individually or could choose to work with others.

Although students were asked how many oranges Rudy had altogether as of square seven, Ms. Benedict was surprised that none of her students gave a total number but rather all of the students told how many oranges appeared when he stepped on the seventh square. Most gave the number as 128 but some of the students used the add-two strategy mentioned above.

It was clear from both groups that some of the students were still developing their concept of just how large a number there would be on square nine. Of those who said that on square nine Rudy would have 512 oranges, one student said that he ate half and then could carry the rest in his backpack, not concerned that he'd have to eat 256 oranges or that he'd have to put 256 oranges in a backpack that would hold only 20.

One student said the following. "When Rudy gets on the ninth square there will be millions of oranges. When I mean millions, I mean millions. Like when I mean millions I mean 512. Then he bends over to pick up the oranges and his back goes out." Another student said Rudy "sold 500 oranges and got approximately $20,823,250,535 dollars." That meant he received about $40 million per orange.

Ms. Benedict commented on what the stories told her. "I feel that the most powerful part of this activity besides seeing and working with the pattern of increasing oranges on each square is the creative writing which incorporates the math. It really helps me to see how the students are thinking about larger numbers."

Possible Extensions

Some interesting extensions or follow-up questions arise from this problem. The first one was posed by a teacher when a group of teachers solved this problem in a workshop setting. "How high would the stack of oranges be on the 9th square?" One group of teachers was so focused on that problem, that they didn't solve the problem as given, but rather explored the space that the oranges would take up. That problem is as valuable a mathematics exploration as the original problem.

What Are the Students Telling Us?

One can observe from the results above that students were engaged at several levels.

It was clear that students' concepts of large numbers were still developing. It seemed that students focused either on the problem or the story, but rarely on both. Overall, students had a range of successful interactions with the problem.

20 DARTS, ANYONE?

The Problem

Ted and Kate were making a dart game for their children. They wanted to have three rings marked with number values. Ted made a board like the one shown. When three darts are thrown, what is the lowest score possible? What is the highest score possible? How many different point totals are possible using three darts?

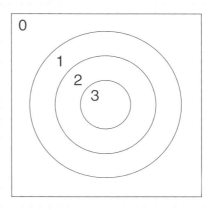

Kate did not like this dartboard because it permitted only ten point totals. That is, using three darts, a person could score 0, 1, 2, 3, 4, 5, 6, 7, 8, or 9 points. Show how each total can be obtained with three darts. Kate wondered whether it was possible to label the rings so that a player could reach a higher score with three darts. She wondered whether all the numbers from 1 to 10—or even higher—were possible. What suggestions would you give for labeling the rings?

Range of Possible Teacher Questions

- What strategies can you use to make the larger numbers?
- Is there only one way to generate a number?
- Can you explain your thinking as you generate other scores?
- Can you come up with any extensions to the problem?

Where's the Math?

This problem has many mathematical facets. It involves mental computation, flexibility with numbers, and logical reasoning. Children may also have to struggle with the notion that if rings are labeled 1, 4, and 7, more values may be obtained but that the longest string of consecutive numbers starting at 1 is from 1 to 9, just as with Ted's original board.

SAMPLES OF HOW THE PROBLEM WAS SOLVED BY OTHER CHILDREN

Ted and Kate were making a dart game for their children. They wanted to have three rings marked with number values. Ted made a board like the one shown. When three darts are thrown, what is the lowest score possible? What is the highest score possible? How many different point totals are possible using three darts?

Kate did not like this dart board because it permitted only ten point totals. That is, using three darts, a person could score 0, 1, 2, 3, 4, 5, 6, 7, 8, or 9 points. Show how each total can be obtained with three darts.

Kate wondered whether it was possible to label the rings so that a player could reach a higher score with three darts. She wondered whether all the numbers from 1 to 10—or even higher—were possible. What suggestions would you give for labeling the rings?

The student work submitted for "Darts, Anyone?" demonstrates that children frequently need work on the understanding-the-problem phase of problem solving. Although it was easy for students to determine what scores were possible on the given dart board, clarification and discussions were necessary for them to understand whether it was possible to design the board to allow scores greater than 9 and still be able to obtain all the numbers, beginning at 1.

Most of the students in Gina Simpson's fifth-grade, Kathy Pickett's fourth-grade, and Amy Benedict's sixth-grade classes were able to show that the given dart board would allow all the scores from 0 to 9 (fig. 20.1). Furthermore, many students showed multiple ways that a sum could be obtained. As might be expected, a few students did not adhere to the three-darts constraint. Other students either ignored the constraints of the problem or just made up number combinations that added to a given number. For example, some students with the numbers 0, 1, 5, 8 on the dart board showed 4 + 4 + 0 = 8 as a way to obtain 8. (See fig. 20.2.)

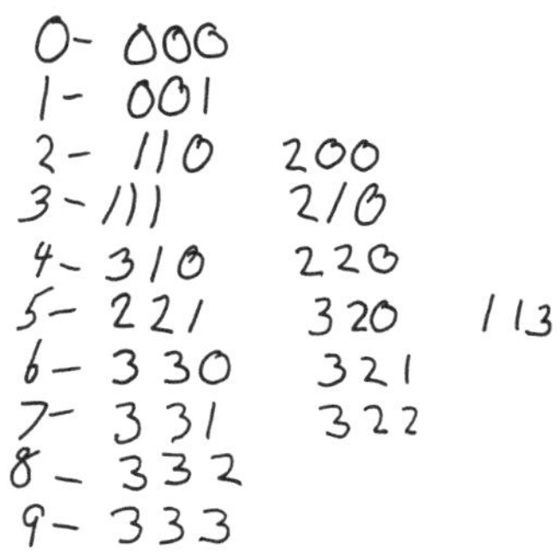

Figure 20.1. A student shows all the possible scores on the original dart board.

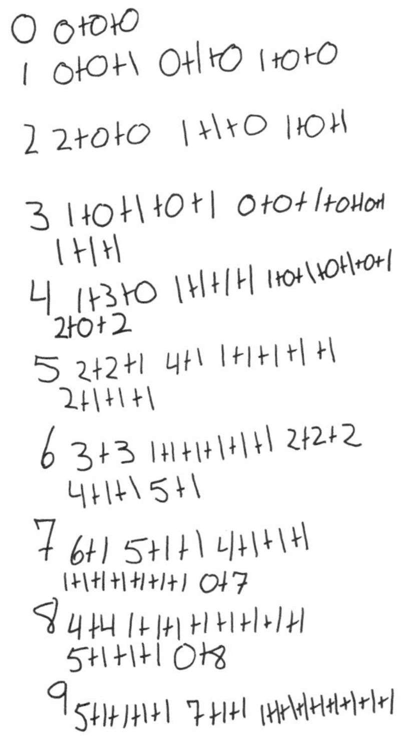

Figure 20.2. Some students did not pay attention to the problem's conditions.

Benedict noted, "Most students could verbalize why the 1 was needed." Knowing that a 0 region and a region with 1 are necessary means that only two more numbers can be used to find a "better" arrangement than the one given.

Benedict also stated, "I asked the students to write a note to Kate suggesting numbers to use and why she should use them." (See fig. 20.3.) Although the students did not produce a full list of possible dart boards, many students did find other combinations that were better than the one given. (See fig. 20.4.) The fact that students were able to find several possible solutions demonstrates that they are able to pursue alternatives to the problem. However, no individual or group was able to show all the possible solutions and provide a rationale.

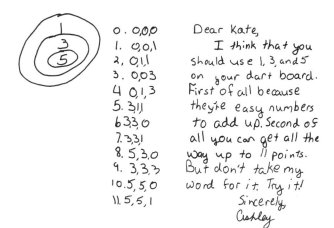

Figure 20.3. Students drew and wrote about their alternative dart boards.

A brief description of the reasoning to find a better dart board follows. As Benedict's students noted, finding all possible solutions is made easier by noting that 0 and 1 must be used. In addition, we can also discount possibilities in which numbers repeat, for instance, (0, 1, 1, 3), since a repeated number does not add potential scores.

To get a three-dart score of 1, one of the four numbers must be 1 and another must be 0. With these two numbers alone, scores of 0, 1, 2, and 3 are possible. To get a score of 4, we must include a number greater than 1 but not greater than 4. Thus, we have three examples to consider: $(1, 2, 3, x)$, $(0, 1, 3, x)$, and $(0, 1, 4, x)$.

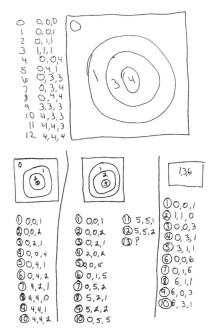

Figure 20.4. Students found several options that gave scores greater than 9.

In the first example, $(0, 1, 2, x)$, scores of 0, 1, 2, 3, 4, and 5 are possible without the fourth number. To get a score of 6, the fourth number must be greater than 2 but not greater than 6, resulting in four possibilities: (0, 1, 2, 3), (0, 1, 2, 4), (0, 1, 2, 5), and (0, 1, 2, 6).

The second example, $(0, 1, 3, x)$, allows scores of 0, 1, 2, 3, 4, 5, 6, and 7 without the fourth number. To get a score of 8, the fourth number must be greater than 3 and not greater than 8, resulting in five possibilities: (0, 1, 3, 4), (0, 1, 3, 5), (0, 1, 3, 6), (0, 1, 3, 7), and (0, 1, 3, 8).

The third example, $(0, 1, 4, x)$, allows scores of 0, 1, 2, 3, 4, 5, and 6 without the fourth number. To get a score of 7, the fourth number must be greater than 4 and not greater than 7. Note that (0, 1, 4, 2) and (0, 1, 4, 3) have been considered previously. This third example yields three possibilities: (0, 1, 4, 5), (0, 1, 4, 6), and (0, 1, 4, 7).

A total of twelve possibilities must be considered. One of these, (0, 1, 2, 3), is the same as the board given in the problem. One possibility, (0, 1, 4, 7), yields scores of 0 through 9, so it is not better than the original dart board. The other ten possibilities are all better than the original board. The possibilities and their scores are listed in table 20.1. Although no one individual or class found or gave an analysis of all possible dart-board configurations with four values, most of the better choices were given in the totality of work submitted. However, no student demonstrated the "best" solution of (0, 1, 4, 6).

Table 20.1
The Ten "Better" Dart Boards

Numbers Used	Posible Scores
0, 1, 2, 4	0–10
0, 1, 2, 5	0–12
0, 1, 2, 6	0–10
0, 1, 3, 4	0–12
0, 1, 3, 5	0–11
0, 1, 3, 6	0–10
0, 1, 3, 7	0–11
0, 1, 3, 8	0–12
0, 1, 4, 5	0–11
0, 1, 4, 6	0–14

What Are the Students Telling Us?

The work discussed here shows that students can become engaged with problems that have more than one possible solution. Students did not need guidance or practice when organizing their work to verify all of the possible solutions. They can achieve success with the problem even if they don't come up with all possible solutions. Students do need time to discuss the problem in order to develop understanding.

21
HOW MANY TOWERS?

The Problem

Herta and Vivienne were excited on their first day of school, especially when their teachers gave them some colored cubes and told them to make up a question to explore. Even though they only had two different colors of cubes—purple and gold—Herta and Vivienne started making towers. First they made towers that were two cubes tall, then three cubes tall and four cubes tall. They began to wonder just how many different towers they could build using only those two colors.

How many towers can you build that are two cubes tall when you have cubes of only two colors to use? How can you be sure that you have made all of the towers? Continue to work with only two colors of cubes. How many different towers can you build that are three cubes tall? Four cubes tall? How can you be sure that you have made all the towers? What patterns do you notice?

Range of Possible Teacher Questions

- Are you finding any patterns for making your towers?
- How do you know if you have all the towers?
- What relationships can you see by studying the towers you've already built?
- Are there any strategies you can use to help you generate towers?

Where's the Math?

In solving this problem, students will be engaged in several mathematical activities. They will be looking for patterns. They will be determining how many combinations can be generated from a limited number of colors. They will be involved in problem solving and communicating mathematically. They will also be extending what they learned from the towers that are two high to the ones that are 3- and 4-cube towers.

SAMPLES OF HOW THE PROBLEM WAS SOLVED BY OTHER CHILDREN

A GREAt deal of work was submitted for this problem. Although most of the solutions came from students in grades 2–6, the work ranged from that of a five-year-old child to that of a high school special education class. In nearly all the responses, children were able to find all the stacks that were two or three cubes tall and made thoughtful predictions for stacks that would be four cubes tall. Students were able to organize their work for three cubes and saw a "mirror" pattern at work. In other words, when they found the stack in figure 21.1a, most quickly saw that the stack in figure 21.1b was also possible. This logic was useful for investigating stacks of four, as well.

Sherial McKinney, Sandra Teel, and Pam Jefferson each posed this problem to second-grade students. Teel and McKinney had the students work in pairs to build stacks of two cubes and three cubes and to predict what would happen for stacks that would be taller than three cubes. Although students used excellent logic to make all the possible towers of three, they had difficulty writing about their thinking (see fig. 21.2).

Teel reported that five of her eight groups of students found all the three-cube towers. Only one group found all 16 four-cube towers. The work of Zachary and Erica shows that they used the mirror method to find all the towers (see fig. 21.3). Although adults might use a different approach to organize their work, this method shows excellent thinking and organizational skills for students in second grade.

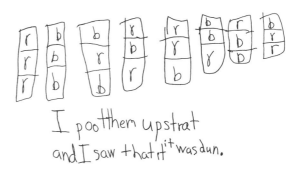

Figure 21.2. Caleb had difficulty explaining how he found the three-cube towers.

Jefferson wrote in detail about the activity:

My second-grade class had previously done a similar problem using three colors of cubes and building towers three cubes tall. I posed the tower problem using two colors and still building towers three cubes tall, and I asked my students whether there would be fewer, the same number, or more solutions. All the students said fewer. I asked why, and one student said, "Because there's not as much to choose from to make them." They estimated that they would find from 10 to 20 towers. After giving out the cubes, all 13 students had found the eight towers within about two minutes. Some worked randomly, but several students said they used some kind of patterning or pairing. Next, I asked how many two-cube towers they could build using two colors. They all knew they would find even fewer. They found the four towers very quickly and were very confident that they had all of them. I then asked how many towers one cube tall could be made. Some immediately answered 2. I wrote on the chalkboard the information we had generated. None of the students saw a pattern in the results until I said the numbers out loud. One student said, "The first one is 2, and 2 + 2 is 4. The next one is 4, and 4 + 4 is 8." I then asked how many four-cube towers we could make. At least half the class wrote down the answer 16. I then asked about five-cube towers, and Kollin said 32. Most of the other students were not sure how to add 16 + 16. Kollin explained that he added 10 + 10 and that was 20, then he added 6 + 6 and that was 12, then he added 20 + 12 and got 32. I pushed even farther and asked about six-cube towers. Over half the students said 64. They could all add 32 + 32 with no problem. As a culminating activity, I read aloud the story *One Grain of Rice: A Mathematical Folktale*, by Demi [New York: Scholastic, 1997].

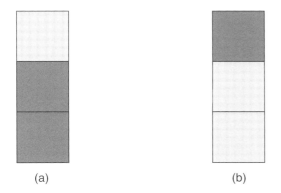

(a)   (b)

Figure 21.1. Students used a mirror pattern to help find all the towers.

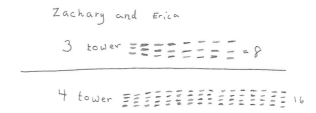

Figure 21.3. Zachary and Erica used the mirror method to find towers.

Nicole Labs guided a group of fourth-grade students to make towers of two, after which they worked in pairs to build three-cube and four-cube towers. Labs reported—

> The students were able to explain their reasoning better as they did each problem. When they first started, they weren't sure of their answers. By the time each pair got to the towers of four, their communication improved and had less dependency on me. I heard more discussion between the children. They would tell each other what they did and why there were or were not more towers. Some examples I heard were "I don't have an opposite tower for this one" and "There has to be more towers, let's look for a pattern." Their organizational strategies did help them. They were able to see when a tower was missing if it didn't have an opposite with it. This strategy hindered one group, because they were missing a tower and its opposite. Each group convinced themselves they had all of the towers by spending time making towers and the opposites. Then, each group checked with other classmates to see how many towers they came up with.

After doing the towers of two, three, and four, all but one pair of students organized their information and were able to extend the problem to towers of five and six. They found that the number of towers that could be made doubled each time.

Andrea McAndrews reported that in her sixth-grade class, "One group of partners immediately saw that the numbers produced were powers of two and wrote, '[The] number of towers would be two to a power where the power is the number of cubes tall that the tower is.'"

What Are the Students Telling Us?

What can we learn from the work of these students? We find that a problem can be investigated at many grade and developmental levels. Even children in second grade can generate solutions and make predictions about these events. Children may not always reason the same way that adults do. But adults don't all reason the same way. Children often look for patterns while they work.

Sometimes students need support in generating different strategies when they find that their predictions are incorrect. As children grow older, their use of strategies to solve problems and their ability to explain their reasoning continues to mature. Problems such as these help in the growth process.

22
THE TALKING CLUB

The Problem

Eric's first-grade class made telephones out of string and juice cans. Students worked in groups and created a telephone club that would connect each person to every other person in the group. If a group had four people, how many strings would be needed to connect every member of the group directly to every other member of the group? The Talking Club, as they called themselves, ended up having a total of twenty-eight strings connecting members in the group. How many people were in the Talking Club?

Range of Possible Teacher Questions

- How many cans and how much string do you need to build a model?
- How did you plan your model so that no one was left out?
- What patterns helped you build your model?
- How would you go about adding a person to your model?

Where's the Math?

Patterns are inherent in this problem. Before students can find a solution, they must become comfortable with the pattern of how the number of strings grows with the number of people. This problem's solution process is similar to that of the well-known handshake problem. In creating and sharing a variety of models that are useful in solving the problem, children explore different representations of the problem.

SAMPLES OF HOW THE PROBLEM WAS SOLVED

TEACHERS used a range of approaches in presenting the problem to their classes. One teacher gave minimal input to the class and let the children collaborate in small groups to solve the problem. During a three-day session around this problem, one pair of second-grade teachers spent much of the first day engaged in conversation with the class about what the group thought the problem was about, making meaning of the task. One group of thirteen K–8 teachers met to discuss the problem prior to using it with their classes. Then they came back together as a group to share their results. One of the teachers in the K–8 group shared how (s)he prompted her sixth graders to look for a pattern and set up a people and strings chart to organize themselves on the second part of the problem. Another teacher began the lesson with a conversation about how cup and string telephones work. The group discussed several questions that arose such as, "Does it matter how many cans there are?"

The approaches used in solving the problem were varied. Some students created models in groups, using cups and string. This approach became frustrating when 28 strings were used. Others used charts like the one below.

Number of People	Number of Strings
2	1
3	3
4	6
5	10
6	15
7	21
8	28

Several students had success with the problem on a variety of levels. Some were able to model the club's number of strings when the number of members was given. Others went on to figure out how many members were in the club if the number of strings was 28. Some students went on to explore ways to generate further numbers from the pattern that they found of adding one more to each successive number of strings than was added to the last. For example, in the table above the number of strings increases first by two from a club with two members to a club with three members, and then increases by one more with each successive number of club members. Still others, including some fourth and some sixth graders found the rule for generating the number of strings for any number of people in the club, where P = number of people and S = number of strings.

$$S = P(P-1)/2$$

Some students who worked individually did not appear to discuss results. They gave their own results and seemed to end the task at that point. Others, like the second graders at Golden Hill Elementary School in Florida, New York, spent several days discussing and solving the problem. The children made decisions ranging from what materials they needed in order to solve the problem to when they needed to switch from one form of model to another. The understanding they showed in their writing at the conclusion of the task was very exciting.

Interesting and Surprising Aspects of the Experience

Elizabeth Halsey-Sproul described how she and Rhonda Schubert used a push-in situation as an opportunity to solve this problem with the second graders at Golden Hill Elementary School. The children worked in heterogeneous cooperative learning groups to solve the problem. As mentioned above, the children were the decision-makers during the problem-solving process. They negotiated many important issues including how many cups the group needed to solve the first problem. The five groups of students chose the following number of cups to solve the problem, by group.

Group:	Number of Children:	Number of Cups:
A	5	5, then 4
B	4	22
C	4	16
D	4	4
E	4	4

Group A originally chose five in order for each person to have one, then later decided they'd need only four and that the fifth person could be the group recorded rather than a model. Group B chose a cup to represent each person in the class. Group C chose 4×4, thinking that 16 showed the number of hook-ups in the group. Groups D and E chose four, but reserved the right to return for more if they needed.

It was interesting to the teachers that the students switched strategies to solve the problem when the numbers became unwieldy with 28 strings. They were discouraged that some students quickly "gave up" when the model with 28 strings became tangled, rather than taking the initiative to find another approach. With the teachers' help they refocused on the problem. Halsey-Sproul writes, "The most troubling aspect of this experience was when the students asked the teacher if what they did was really math because they spent most of their time talking and thinking and not adding, subtracting, or multiplying."

What was especially exciting to the teachers in this class was the fact that a push-in student took on a strong leadership role in her group and provided key information needed to solve the problem. It is another example of the benefit of communicating mathematically. When children are allowed to collaborate and share their understanding as part of a community, often the children who are considered "weak" by traditional standards have opportunities to show the strength of their understanding.

Overall, in the process of solving the "Talking Club" problem students generated questions, discussed strategies, and built models that led them to visualizing the pattern. Strategies included acting out, drawing pictures, using geoboards, and constructing charts for organizing data. Different forms of representations of the situation gave students different insights into the problem.

What Are the Students Telling Us?

This was a rich problem in that there were multiple avenues toward the solution, but also teachers were able to create other, related problems, such as the Ping Pong Tournament, to extend student thinking. It was interesting to note that in solving this problem, students communicated with other students, and teachers collaborated with other teachers as they analyzed and reflected on the successes of the students. A culture for learning had been established in the mathematics classroom.

References

Devine, Donald, Judith Olson, and Melfried Olson, *Elementary Mathematics for Teachers*, 2nd. ed. New York: Wiley & Sons, 1991.

23
FAIR SHARE

The Problem

The Noslo twins, Tanya and Travis received a birthday cake from their Aunt Colleen. Aunt Colleen is a mathematics teacher, so she made the cake in a special shape, a 2 × 3 rectangle, with markings like a geoboard. The twins decided to make a drawing of the cake on a geoboard and use one rubber band too cut the cake into exactly two pieces so that they would each get a fair share. In how many different ways can they divide the cake?

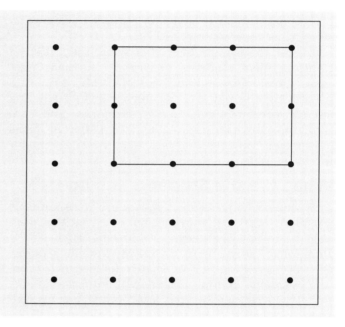

Range of Possible Teacher Questions

- What do you think they mean by "fair share"?
- Is there more than one way to cut the cake?
- Can you categorize the types of cuts you're making?
- How do you know your cuts create fair shares?
- Can you come up with any extensions to the problem?

Where's the Math?

While searching for solutions, the students will be involved with the mathematical ideas of area, congruence, partitioning, and fractional pieces. To find all the solutions the students must also develop a logical approach or a classification scheme. The students will have to decide whether solutions that are flips or rotations of each other are different. This decision will determine how many different solutions they count.

SAMPLES OF HOW THE PROBLEM WAS SOLVED

STUDENT solutions to this problem showed a range of thinking. Almost all the students in Courtney Veronesi's third-grade class generated at least five solutions while working with a parent helper. Almost all the solutions involved congruent pieces. Linda Sandersen presented this problem to third-grade Title I students. Most of them were able to make at least one fair division, but "some had trouble using only one rubber band for partitioning." Sanderson could not discern many strategies in her students' work, and her students did not explain that their fair shares had the same area.

The students in Angela Andrews's fourth- and fifth-grade mathematics class were engaged in exploring the problem for two days and generated wonderful ideas in the process. Andrews presented the problem without providing guidance about congruence or area so that she could see how her students would do on their own. A common early error was forgetting that the cake could have only two pieces. The students discussed flips and rotations of cutting schemes and whether these divisions should count as different: "The class came to a group decision that if it can be flipped or rotated to be congruent to a way already shown, it will not count as a new way."

While observing the students at work, Andrews noted that those students who began their exploration using dot paper first were involved in less discussion and made more errors. Those who worked with geoboards first engaged in more discussion and produced examples that fit the structure of the problem better. Andrews asked students to justify their reasoning; when they began using visual arguments, she "pressed them for a more sophisticated rationale by reminding them that many things might look fair, but their job was to use mathematics to prove their ideas. The children then began to analyze their examples in terms of the relationships between the two parts." As a result, students used such ideas as symmetry, congruence, and area to explain their decisions. The children also concluded, "If the shapes are congruent, they have to be fair shares, but fair shares don't have to be congruent." Of course, some of the students' generalizations were incorrect, especially those that involved the perimeter of pieces. Through discussion, students were able to conclude that pieces with the same perimeter might not have the same area. Through this intense investigation, the class was able to generate almost all the possibilities.

The students in Kristina Kalb's fifth-grade class did generate all the solutions. Of course, as the students in both Andrews's and Kalb's classes indicated, the number of solutions varies according to what configurations are counted as different. Both classes thought that if the figures were congruent, they should not count as different. Kalb's students produced the nineteen examples shown in figure 23.1. In describing their solutions, these students agreed with Andrews's class, saying, "We only counted box number 1 and not box 1B because it is the same cut just turned on the vertical axis. 2 and 2B are horizontal axis flips from each other, and so are really the same thing. The following pairs are the same because:

    3, 3B—horizontal or vertical flip
    4, 4B—horizontal flip
    4, 4C—vertical flip
    4, 4D—horizontal flip, then vertical flip
    5, 5B—vertical flip
    6, 6B—horizontal or vertical flip
    7, 7B—horizontal or vertical flip
    8—has no other flips
    9, 9B—vertical flip"

(The last solution was the only example not generated by Andrews's students.)

This work shows that children can deal with mathematical concepts beyond the visual level. These students made good use of mathematical language and concepts to explain their reasoning. They discussed fairness and determined that equal areas would be needed. They also discussed the idea that figures that produced congruent results should not be counted as different.

Andrews elaborated on her students' reasoning about this problem. Her students thought that the shape shown in figure 23.2 represented a fair share because the directions did not say that the band that cut the cake had to touch the nails. This solution generated a new problem to examine. Many students believed that if the band did not have to touch the nails, then an infinite number of solutions was possible.

Andrews also reported, "One child dropped the rubber band onto the board and pondered if there would be a way to cut a circle out of the cake and make fair shares. The class decided that 'We'd have to have a cookie cutter with an area of 3 square units.'" Finding a way to create that circle was difficult because the students had not studied the relationship of the radius and area of a circle. This discussion was soon followed by another when a student wondered whether a rectangle could be cut out of the cake to make fair shares: "After much calculation, the group came up with the solution that a rectangle 2 units by 1 1/2 units would work." The class then wondered whether a square could be cut out to make fair shares. They found that the answer is yes, but they were unable to determine the length of the side for a square that has an area of three square units.

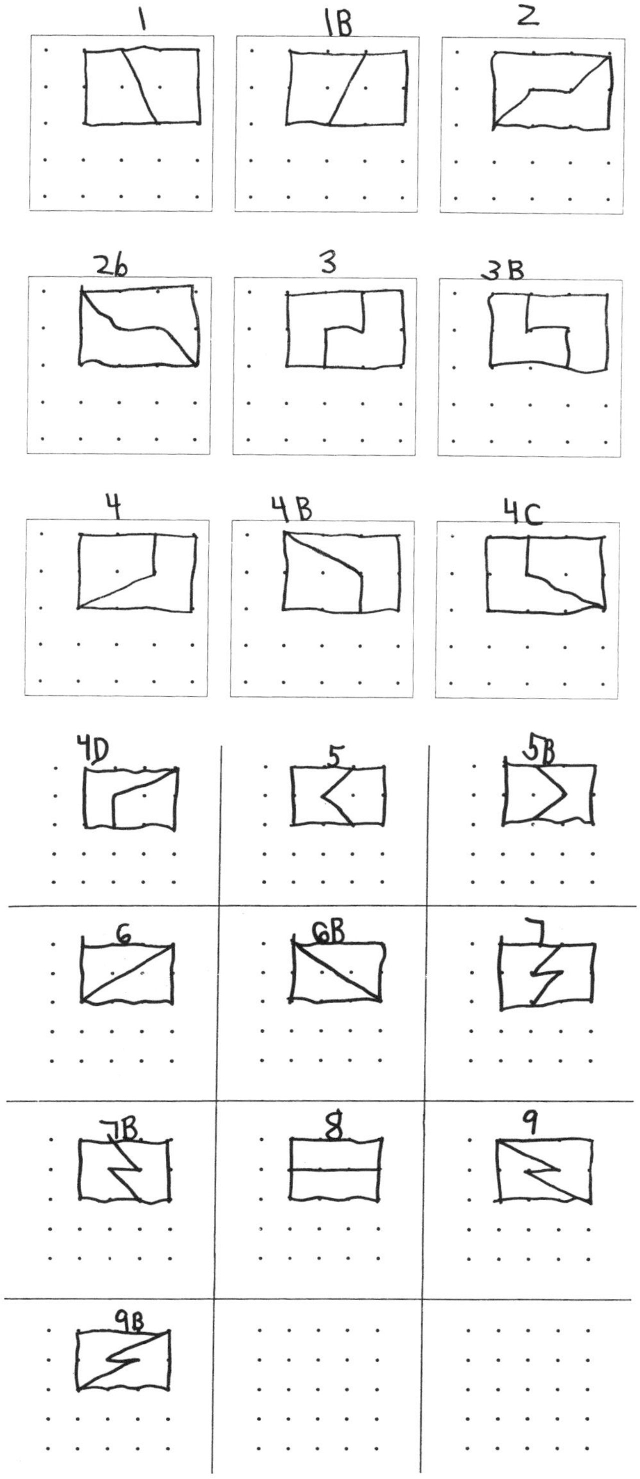

What Are the Students Telling Us?

The students showed once again that they can tackle explorations in which they generate and wrestle with significant mathematical ideas. When the students worked outside the more obvious parameters of the problem, they came up with interesting ideas to explore.

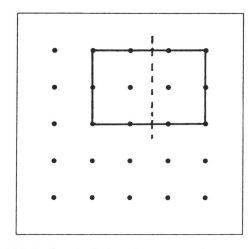

Figure 23.2. Students' solutions that generated a new problem

Figure 23.1. Nineteen solutions from one fifth-grade class

24 MATHEMATICS OF MOTION

The Problem

Variation 1. Using the design shown, find a series of moves to take the design from the top-left corner of a piece of paper to the bottom-right corner, in its original orientation. A move must be a flip, a turn, or a slide; and you must use more flips and turns than you use slides. How many moves were needed? Draw each move on a piece of paper, or use pattern blocks to build a model of what the design looks like after each move. Try to do the task again in fewer moves. What strategies helped you do so?

Variation 2. Write instructions to describe your series of moves. See if another student can recreate your moves by using only the instructions. Could a different series of moves be used to achieve the same result?

Variation 3. Using Logo MicroWorlds from Logo Computer Systems, Incorporated (LCSI), recreate the design as a stamp. Make it into a turtle, and move it through any of the activities suggested here, using the Logo commands and the mirror-image capabilities of the stamp creator. You might want to leave a stamp of the design at each position to show your path.

Variation 4. On a playground, design an obstacle course in which your classmates use slides, flips, and turns to move from a starting point to a finishing point. Write a careful description of how they must move to complete the course.

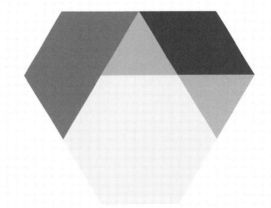

Range of Possible Teacher Questions

- What relationships can you see by studying the moves?
- Which moves are easiest to make? To describe?
- What are the different ways we can describe a turn? A flip?
- Are there any strategies you can use to help you decrease the number of moves needed?

Where's the Math?

"Transformations are important aspects of geometric thinking. Young children come to school with intuitions not only about shapes but also about how shapes might move. Through school experiences with informal motions such as 'slide,' 'flip,' and 'turn' made by using mirrors, paper-folding, and tracing, elementary school students can see these ideas as mathematical in nature" (NCTM 1998, 61–62). The explorations with the problems discussed here allow children to create a variety of representations that they can use to talk about how objects, including themselves, move through space. The problem offers many entry and exit levels to allow children of varying abilities to be engaged in exploring the mathematics of motion. Additionally, it allows children to explore the range of positions through which the objects pass in the process of repositioning. The language used to describe such changes is rich, and the problem gives children the opportunity to explore and challenge their own and others' understanding of spatial sense. They also have an opportunity to talk and write about how objects move through space.

SAMPLES OF HOW THE PROBLEM WAS SOLVED

VINNY Carbone, a fourth-grade teacher from Fairfield, Connecticut, tried this problem with his mathematics class. Before introducing the problem, he designed activities to teach his students the meaning of the terms *turn*, *flip*, and *slide*. To demonstrate a turn, he used a square tile and traced the image on the chalkboard after rotating it 90 degrees. Then he asked the students to move the square from the upper left to the bottom right of the paper.

As Carbone observed the class, he noticed that certain students did not keep one corner stationary as they turned the square tile. To demonstrate a flip, he used a mirror to help students visualize the reflected image and to answer their questions concerning whether it was okay to flip the image on its corner or upward, right, or left. One student asked whether the square could be upright on the paper. Another student devised a game with these rules: "The board contains all squares. You roll the dice, and whatever number you get, you can make that many flips. The object is to get into the bottom square first. If your square ends up on an opponent's tile, the opponent has to go back to the beginning." At the end of the day, Carbone reflected that he wished he had used a rectangle rather than a square to introduce the activity.

On the second day, Carbone described a slide and then asked students to solve the stated problem using the design shown in the problem. One interesting aspect of this activity was that students asked lots of questions in an effort to clarify their thinking about the problem and the concepts. Some students were confused about the restriction concerning the number of flips and turns compared with the number of slides, and others were confused about being able to trace over areas already used. Some asked conceptual questions, such as whether they could turn on a trapezoid's corner or whether they could make a flip and a turn on one move.

Carbone made some observations that might be helpful to those who try this problem with their classes:

- Color coding their trace marks helps students keep track of each piece.
- Students are most comfortable with flips and slides; only a few tried a turn.
- Students at times were doing flips and calling them "turns."
- Many students would rotate a pattern block only about the center of the piece.
- The students and teacher were most challenged by rotating the hexagon, rhombus, and trapezoid through 90 degrees.
- In the flipping process, the thickness of the pattern blocks caused some students to leave a space between the figure and the image.

What Are the Students Telling Us?

As a result of working on this problem, students developed a clearer understanding of how to visually represent three important transformations. By asking questions and sharing their perceptions, the students began to clarify their own thinking about centers of turns, lines of reflection, shapes of images, and properties of various shapes. They were also building an intuitive sense of what features of a given shape are preserved under certain transformations. The teacher, as facilitator in this class, was able to benefit by reflecting on his own teaching. He indicated that at times, he "should have pressed further and asked [the student] his thoughts. 'Why do you think this is right?' 'What do you think is wrong with it?' 'Something seems to bother you about this picture; what does it seem to be?'" This problem-solving experience created an opportunity for both teacher and students to grow in their ability to understand and communicate mathematics.

25
MARBLES

The Problem

Sheryl was excited when she arrived home with three small boxes of marbles. She labeled each box with its contents. One box had 2 blue marbles, a second box had 2 red marbles, and the third box had 1 blue and 1 red marble. The next morning, she found that her mischievous little sister Angelica had played a trick on her. Angelica had removed all the labels and placed them back on the boxes so that each box was labeled incorrectly.

Sheryl wanted to put the correct labels back on without opening all the boxes. She wondered if she could figure out the correct labeling by just seeing the color of 1 marble in one box. Is it possible for Sheryl to pull 1 marble from one box and know the correct labels for each of the boxes? If so, how? If not, why not?

Range of Possible Teacher Questions

- What does pulling a blue marble from the box labeled Red tell you?
- How many possible ways of starting are there?
- What kinds of models can you use?
- Are there any strategies you can use to help you determine relationships?

Where's the Math?

This problem will engage students' logical reasoning and should raise such questions as "What if I try this?" and "What do I know now?" Some students will be able to visualize and reason about what could be in the boxes, whereas other students may need help in developing a strategy.

SAMPLES OF HOW THE PROBLEM WAS SOLVED

The responses from the students in Brenda Deihl's class lend wonderful insights into how sixth-grade children tackled this problem. Their work demonstrates how they struggled to understand the problem, reasoned about and analyzed the possibilities, communicated ideas, and reflected on their work and thinking processes. Deihl structured the assignment to indicate that she expected explanations of the students' work along with their solutions. She presented the problem as it appeared in the journal and gave the students the assignment shown in figure 25.1. She allotted "two weeks to complete the problem outside of the classroom. Once a week, we discussed any problems or difficulties that the children encountered." Not only did her format provide guidance for the work that she expected, but the schedule also allowed the students time to reflect on their thinking.

As we might expect, the responses from students varied. The writing submitted revealed that some students thought that they were finished if they could identify one possible way to solve the problem. Others realized that they must guarantee that Sheryl would know the correct labeling without a doubt after only one draw. The solution processes and reflective thinking demonstrated by these students are remarkable.

Some students thought that perhaps Sheryl could remember a location or recall the color of one box and, thus, would be able to figure the problem out. Dan shared this opinion (see fig. 25.2), but he also reasoned that for some choices, Sheryl would not be able to tell the boxes apart with certainty. Dan demonstrated that Sheryl cannot just randomly pick a box, but his reasoning is incomplete. A few other students offered reasoning similar to Dan's.

MARBLES

Your task:

1. Decide if it is possible for Sheryl to pull 1 marble from one box and know the correct labels for each of the boxes.
2. Design a clear way of displaying your mathematical reasoning for each possibility that you considered.
3. Include a written explanation as to how you chose the outcome that you thought was correct and eliminated the incorrect choices.

Figure 25.1. Brenda Deihl's assignment of the problem

Sheryl can figure out what marbles are in what box by simply remembering what marbles are in what box. A certain mark or scratch on the box would easily give the answer away. She could also remember what order the boxes were in.

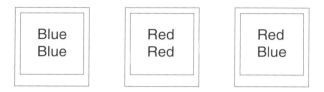

The reason why you couldn't just pick a marble from a box and decide by what color the marble was is because you don't know what color the marble in the box was. If you pick a blue marble out of a box, that narrows your choices down to either blue-blue or blue-red. If you pick out a red marble, that narrows your choices down to either red-red or red-blue. However, there is no way to determine whether or not it is all one color or a mix of the two.

Figure 25.2. Dan's solution

Several students examined solutions that would allow Sheryl to identify the boxes by selecting only one marble from a box, but they did not fully explore the possibilities. Audrey and Ashley tried to solve the problem by checking the six possibilities that exist for drawing marbles from the boxes. Audrey's work is shown in figure 25.3. She illustrated several situations that would allow Sheryl to find the correct labeling, but her work was not systematic enough to develop a strategy that will always work. Ashley's work was similar to Audrey's, but she went a step further, noting, "The best one with a definite answer would be picking out of a wrong-labeled blue-red cup."

Beth demonstrated her understanding of the problem with a skit about Sheryl and Angelica that was videotaped by her parents. She was quick to say that she came to her solution after "I spent a week sitting around the house saying 'This problem is impossible' and trying to figure out what the answer was." Beth used strips of paper to list all six possible arrangements, then carefully narrowed the number of choices down to the two arrangements that modeled all three boxes labeled incorrectly (fig. 25.4). She then demonstrated that if she drew a blue marble from the box labeled blue-red, she could determine the correct label. A couple of other students used this same reasoning, but they did not show that if a red

Audrey

I believe that Sheryl can pull 1 marble out of 1 box and know the correct placing of the other marbles. Here is my thinking:

The answer to the "Marbles" question involves using the process of elimination and just thinking about the problem. by process of elimination you can accurately predict the marbles in the other boxes. My theory is that since Sheryl's sister labeled <u>ALL</u> of the boxes incorrectly, you can narrow it down to the correct marbles except in two cases.

Color key: purple-sister's labeling green-correct labeling
Leter keys: B=blue R=red
Answer 1:

Say you go to the **RR** box and pull out a red marble. You know that it's definitely Not the **RR** box because her sister labeled ALL of the boxes incorrectly. So you've narrowed it down to the last possible answer - **RB**

Answer 2:

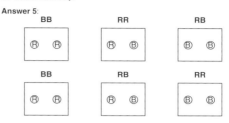

Say you go to the **RB** box and pull out a blue marble. You know that the box is **BB** because sis labeled ALL of the boxes wrong and that is the only other possible answer.

Answer 3:

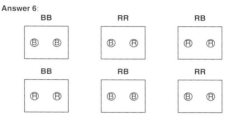

Say you go to the **RB** box a ull out a red marble. You know that the correct labeling for that box is **RR**. It can't be **RB** because all of the labels Sheryl's sister labeled were labeled incorrectly.

Answer 4:

Say you go to the **BB** box and pull out a blue marble. You know that it's the **BR** box. It cannot be the **BB** box because Sheryl's sister labeled ALL of the boxes incorrectly.

Answer 5:

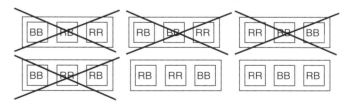

If you pull a blue marble out of the box labeled **RR** you may or may not get the correct placing of the other marbles. You can narrow it down to one of two boxes yet you can't be sure which one it is. You know that the box is either going to be the **BB** or the **BR** as I demonstrated in the diagram. In this case you don't know for sure which box it is.

Answer 6:

If you pull a red marble out of a box labeled **BB** you may or may not get the correct placing of the other marbles. You can narrow it down to one of two boxes but you cannot be positively sure. You know that the box is going to be **RR** or **RB** as I demonstrated in the diagram above. Unfortunately you can't tell in this situation.

Figure 25.3. Audrey's solution

marble was drawn from the box labeled blue-red, the correct labeling could also be determined.

Many students provided diagrams or pictures of their thinking and record keeping as they worked through this problem. Kelly produced both an excellent diagram to demonstrate her solution (fig. 25.5) and a careful description of her thinking (fig. 25.6).

These students did an exceptional job of writing about their ideas and reflecting on their thinking. The importance of the teacher's role is shown in the structure of the assignment and the requirement that students write about their thinking processes. Several students clearly described their frustrations, as well as the turning points in their thinking that allowed them to be successful. For example, Ashley said, "At first I never thought that this problem would work at all. This was because I wasn't looking at all the facts, I was only looking that there were three boxes and she could only pick one marble from one box. I was thinking, no way is this possible; I wasn't thinking too hard. One fact I was missing was that all the labels were put back on incorrectly." Most students began to see a solution strategy when they focused on the idea that all the boxes had incorrect labels.

The richness and creativity of Beth's reasoning is demonstrated in the following scenario.

The boxes above are shown as Beth sees them.

The six strips above show all the possible arrangements of the boxes. Beth then eliminated all the arrangements in which either box 1 is in the first position, box 2 is in the second position, or box 3 is in the third position, because these arrangements would have one of the boxes correctly labeled. She was then left with the two arrangements that showed all three boxes incorrectly labeled.

Figure 25.4. Beth's solution

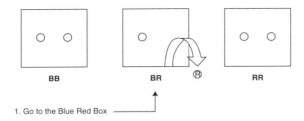

1. Go to the Blue Red Box

If you pulled out a Red marble, then you know that, that box is the **RR**.

You know that, because, the only boxes that would contain a red marble would be the **RR** or the **BR**. And since you picked it out of the **BR** box, then that box must be the **RR** box. (Because they are all incorrect)

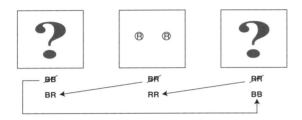

Since the labels are all incorrect, and you know that the **BR** box is really the **RR** box, then you can just switch the other labels, so that the box that was labeled **RR** is the **BB** box, and the box that is labeled **BB** is labeled **BR**.

This is what the inside of the boxes would look like. So yes it is possible for Sheryl to pull one marble, and tell which marbles are in each box.

Figure 25.5. Kelly's diagram of her solution

I think that it is possible to pull one marble from one box, and know the correct labels from each box.

I came upon my answer by just making sense of the information that was given. Example: You know that all of the labels are labeled incorrectly, and you know that two of the boxes contain red marbles; or blue marbles, whichever. So just go to the box labeled blue-red. (You wouldn't go to the box labeled red-red, or blue-blue, because if you were to pull a red from that box labeled blue-blue, or if you were to pull a blue from the box labeled red-red, then the blue or red marble that was pulled out could be from the blue-blue box or the blue-red box, or if it was a red marble pulled out from the blue-blue box, it could be from the red-red box or the blue-red box. So you wouldn't be able to eliminate one of the two boxes.) And if you pulled out a red marble, you would know that that box would have to really be the red-red box. (It would have to be the red-red box because the only boxes that would contain a red marble would be the red-red box or the blue-red box. Since the labels are all incorrect, that box would have to be the red-red box.) The same reasoning applies to if you pulled a blue marble from the blue-red box.

I know that my conclusion is right because it works.

Figure 25.6. Kelly's description of her thinking

One satisfying result of this assignment is the amount and variety of parental involvement mentioned by the students. Many students noted that they worked alongside their parents to model the situation. The students' comments clearly showed that their parents did not know the solution to the problem and thus were equal partners with hteir children in a problem-solving situation.

This work shows that sixth-grade children can be reflective thinkers and work on a problem for a significant amount of time. Children can make conjectures about possible solutions and furnish evidence to prove or reject their conjectures. In doing so, they exhibit their ability to reason and justify their thoughts.

What Are the Students Telling Us?

We saw students struggling to understand the problem at first. This meant that they needed time to discuss and make sense of the problem before trying to solve it. It was exciting when they did reach an understanding and then began to reason about and analyze possibilities. Their communication and reflection about their work was a very important component of the process.

26
WHAT CAN YOU MEASURE?

The Problem

Kudit Strate was a carpenter who worked on counter tops. One day, she came to work without her ruler. However, she located an object made of wood segments with interesting numbers on each segment. She decided that the markings indicated the length of each segment in inches. That is, the longest of the segments was 8 inches. At one end of the 8-inch segment was a 7-inch segment attached by a fitting that allowed the segment to swivel. On the other end of the 8-inch segment were 3-inch and 6-inch segments that would also swivel.

Strate knew that she could use this contraption to measure 3 inches, 6 inches, 7 inches, and 8 inches but wondered whether she could use it to measure other lengths. What other lengths can she measure? Can you design a similar piece of equipment with four segments what would be "better" than the one Strate has?

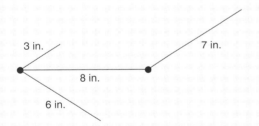

Range of Possible Teacher Questions
- What relationships can you see by studying the diagram?
- Are there any strategies you can use to help you determine relationships?
- What lengths can Strate measure using this object?

Where's the Math?

While searching for solutions to this problem, children will be involved with measurement of length. To find all the solutions, children must develop a logical approach. Children will have to decide how to swivel the pieces to find measurements of other lengths. They will also be practicing adding and subtracting. The investigation to find a better design is open-ended and will require a large amount of thinking, which will include deciding what "better" means.

SAMPLES OF HOW THE PROBLEM WAS SOLVED

THE work of the fourth-grade students from Kathleen Schanbacher's class in Lewisburg, Pennsylvania, presented interesting reading and contemplation. The work lends wonderful insights into the thinking and organizational skills of students. Clearly, the students understood the problem. What is not clear, of course, is what is meant by a "better" piece of equipment. Students need to discuss what *better* might mean. Most thought that having a piece of equipment that would allow for a greater variety of measurements would be better. Some students, however, commented that some configurations might measure angles, some might be easier to carry, and some might measure larger numbers.

Although some students seemed to think that a different configuration might be an option, many of them chose to use numbers that were identical to those presented in the problem (see fig. 26.1). In other words, students could escape one constraint, but many could not escape two constraints. Only two students chose to work with numbers other than 3, 6, 7, and 8 (see fig. 26.2).

Four prevalent themes emerged from the students' work:

1. Students concentrated on seeing how many combinations they could find rather than on making sure that each combination gave a measurement that had not already been found. This focus demonstrates that the students understood the problem but that they did not reflect on all aspects of it.

2. Some students became confused between what measurements they could make and what they would be able to measure with the instrument. For example, some students thought that the arrangement in figure 26.1b could measure an object of length 24 because 24 = 6 + 7 + 8 + 3. Yet the equipment would not allow the 6, 7, 8, and 3 to line up to measure an object of length 24. Such confusion emphasizes the need to discuss the problem after children have had a chance to explore the conditions. When students share their ideas with one another, they become aware of different thoughts that emerge. Then the class can decide whether a certain approach is allowable when working this problem.

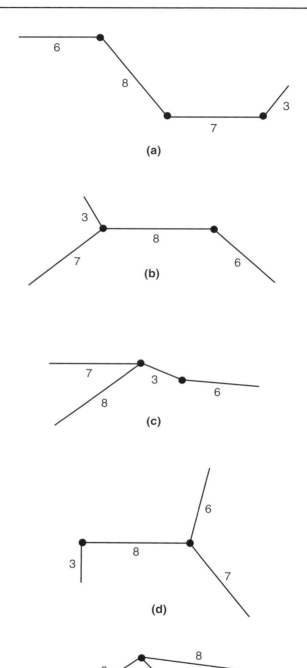

Figure 26.1. Students' configurations using numbers presented in the problem

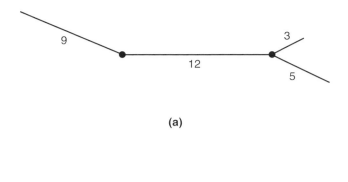

(a)

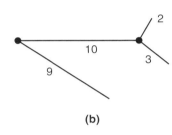

(b)

Figure 26.2. Student's configurations using numbers other than those in the problem.

3. Not surprisingly, most students were unable to find all the possible measurements for their arrangements. Students may have difficulty keeping the configuration in mind while considering the options to add and subtract to determine lengths that could be measured. Many students found arrangements that would generate a greater variety of measurements than Strate's but were not able to verify this accomplishment. For example, the configuration in figure 26.2b allows for measurements of 1, 2, 3, 4, 5, 7, 8, 9, 10, 12, 13, 16, 17, 19, 21, and 22—more measurements than Strate's, which allows for measurements of 1, 2, 3, 4, 5, 6, 7, 8, 9, 11, 12, 14, 15, 18, and 21.

4. Students' organizational abilities are still developing. Only two students showed methods that organized the values obtained. Figure 26.3 shows the work of one student who attempted to list the number of measurements that could be found, but some measurements were repeated. The work in figure 26.4 clearly demonstrates a fairly sophisticated organizational attempt, as well as details of the thinking process.

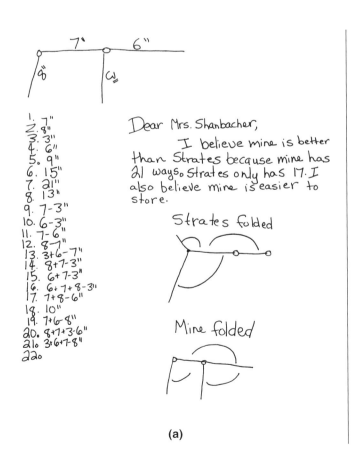

(a)

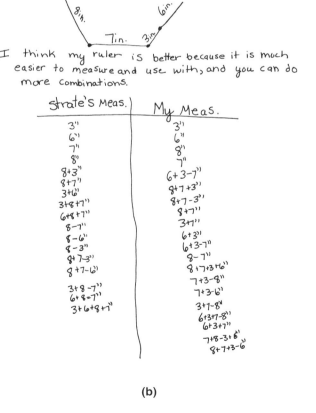

(b)

Figure 26.3. One student attempted to list the number of measurements.

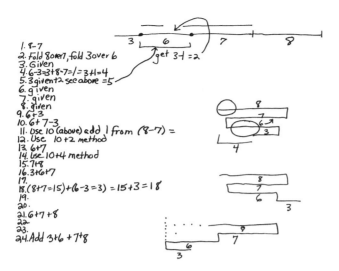

Figure 26.4. This student explained the organized way that the problem was solved.

What Are the Students Telling Us?

Students were clearly engaged in thinking mathematically, yet they needed guidance in organizing ideas and reflecting on work. The variety of arrangements generated by students provides even more opportunities for exploration. For example, students could examine whether "linear arrangement" in figure 26.1b would always generate the same possible measurements, depending on the placement of 3, 6, 7, and 8. this question would be natural to ask because many students used this general arrangement.

27
WHICH GRAPH IS WHICH?

The Problem and Possible Teacher Questions

The two graphs below show information about two different things related to a given class. One shows how many pets each of the thirty students in the class has and the other shows how many times in a month each child in the class brings lunch to school rather than eating the food served in the cafeteria. Which graph do you think represents which data set? Why? When you decide that, label both axes on each graph. Then write a letter to a friend who disagreed with your choice. Explain why you think your representation is correct.

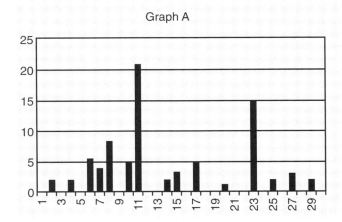

Graph A

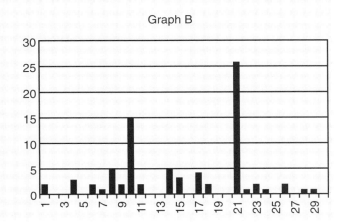

Graph B

Which of the Following Questions Can Be Answered from the Graphs?
- How many students have cats for pets?
- On which day do most students pack a lunch?
- Which student has the most pets?

Choose one of the graphs and describe another data set that the graph could represent. What can be determined from the data in your new representation?

Questions for the Teacher to Reflect On

- What strategies did the children use in determining which graph fit which situation?
- What were common difficulties that children had with the problem?
- How well did students explain and support their choices?

As children interact with data represented in graph form, they have a range of choices about how the information is represented. It is important for children to explore graphing options and determine what one can conclude from a graph as well as what cannot be concluded from the graph. This exploration allows for a range of discussion about what is known about a data set from the information represented, and about what cannot be determined from the graph.

Where's the Mathematics?

This problem represents a range of explorations currently being used to help children develop graph sense. Graph sense is described by Friel, Bright, and Curcio (1998) as the ability to:

- Compose and decompose graphs
- Recognize elements and interrelationships among elements in a graph
- Speak the language of graphs when reasoning about information in graphs
- Describe the shape of the data from the graph
- Recognize when one graph is more useful than another based on knowledge of the data being represented
- Move flexibly among graphs and see interrelationships among graphs
- Respond to different levels of questions associated with graph comprehension

(Friel, Bright, and Curcio 1998)

SAMPLES OF HOW THE PROBLEM WAS SOLVED

THIS problem challenged students to identify what is known about a data set from the information represented. Interpreting the meaning of the tallest bar in each graph in the context of the problem was the key for determining which graph was which. When looking at the bar with the height of 26, some students realized that a person could have 26 pets, but addressing the meaning of 26 as a piece of data in the lunch setting provides a clue for the solution. Although the number of pets that a child could have is unlimited, the number of days that a child can attend school is not. If a child attends school five days per week, then he or she will attend school no more than twenty-three days in a month, and that situation could occur only in the seven months that have thirty-one days each.

Thus, the second graph cannot reflect the number of days children packed lunch.

Gabriell Sacks of Lansdowne Friends School in Lansdowne, Pennsylvania, explored the problem with students in a combined fifth- and sixth-grade class. In recent lessons, the students had examined what Sacks referred to as "mystery" plots that required them to match plots to data sets. Students found this experience challenging. The teacher saw the "Which Graph Is Which?" problem as a logical extension of the work that they had been doing.

The children reasoned using their own experience with pets and with the number of children typically packed lunch in their school. This reasoning reminds us that problems must have a meaningful context for children to make sense of them. Many children focused solely on their experience or solely on reading the data, although a few both interpreted the data and used their experience to make a decision. The children used three different strategies in solving the problem. The first strategy was to read the graph and determine what information could be used to solve the problem. The interpretations of the labeling were varied. Some children described the horizontal axis correctly as the number of students in the class, stating, "It says there are thirty students in the class, and both graphs have the same amount of spaces." Other students described the horizontal axis as "types of pets" on graph A and "students" on graph B. Some students interpreted the fact that graph B is numbered vertically to 30 to mean that the graph must be about the number of lunches packed in a month, because a month has 30 days. Another student read the horizontal axis as representing the names of children on graph A and the names of pets on graph B, although in both instances the data given were represented by numbers. Other students labeled the horizontal axis as representing the number of school lunches per student and the number of pets per student. The reasons that the children decided to label the horizontal axis as they did would make an interesting class discussion.

On the vertical axis, several students recognized that the number represented either how many pets a child had or how many lunches the child brought.

One child labeled the vertical axis on graph A as follows: 5-fries, 10-soda, 15-nuggets, 20- cheeseburger. Graph B was labeled 5-birds, 10- hamsters, 15-dogs, 20-rnice, 25--cats, 30- ferrets. Talking with this student would give the teacher an opportunity to understand his reasoning for labeling graphs in this way.

Students' strategies for labeling were related to how successful they were in solving the problem. The labeling indicated an understanding of how the graph should be interpreted. One reason for their success may be that students who understood how to label and read the graph may have used that knowledge in solving the problem. Alternatively, these students may also have had a better understanding of the problem and thus were more successful in labeling and interpreting the graph.

The second strategy that students used was to relate the problem to real-life experiences with pets and with children who bring lunches to school. For example, one child talked about whether children were allowed to have dogs as the main reason in choosing a graph to represent a data set. Another child noted that about half the children in his class bring lunch every day and that the other half eat cafeteria food. He said that as a result he could easily determine that graph A represented how many lunches a child brings in a month. Talking further with this child about how he concluded that graph A represented half the group bringing lunch each day might prove beneficial. Two of the children combined these first two strategies in solving the problem by using data points and what seemed to them to be reasonable explanations. For example, one child stated, "I think graph A was cafeteria, because it makes sense that kids pack a lunch 21 days of the month or 1 day."

The third strategy that students used was to follow their instincts. One student stated in his written explanation that everyone else in the class had gotten a different result, but instinct told him that graph A should be pets and graph B should be lunches. Again, hearing why that result made sense to this student would be interesting. His statement that his instinct told him he was right is one that we often hear from children. Such an explanation may indicate that these children have not yet articulated the reasons that their answers seem reasonable to themselves; they may be afraid that their results may not be accepted; or as this student suggests, the answer may just seem right. Any of these indications present an opportunity to talk further with the child about his or her reasoning. Sacks's thoughts about the experience of determining Which Graph Is Which? were that the students are still developing an understanding of the meaning and interpretation of graphs and that personal experience played a big role for several of the children. Their mathematical thinking reminds us of the importance of giving students opportunities to solve good problems. Such problems require them to examine real-world data, discuss their thinking, make decisions about the data, and describe how they made those decisions.

What Are the Students Telling Us?

Students told us, through their work, that they are capable of interpreting graphs as well as drawing inferences from them. They also told us from their misconceptions that we need to do more work with developing graph sense. Their engagement with the problem told us that the real-life connection seemed to help them situate the problem in a meaningful way.

References:

Friel, Susan, George Bright, and Frances Curcio. "Graph Sense". Paper presented at the meeting of the Research Council for Diagnostic and Prescriptive Mathematics, College Park, Maryland, 1998.

28
STAIR SKIPPING

The Problem

Alexis, Hamilton, and Dallas were at the bottom of a rickety spiral staircase that was their only way out of a mineshaft. They could leave only by going out one at a time and thought that their best plan was not to step on the same steps on the way out. After a short consultation, they made a plan: each person would step with his or her left foot on the first step. After that, Alexis would walk the steps one at a time, alternating feet as usual. Hamilton and Dallas would also alternate feet as usual, but Hamilton would skip one step and Dallas would skip two steps on their way out. They pondered a few questions:
- On which step from the bottom would their right feet first land?
- On which steps would all three people land?
- On which steps would they all tread with their right feet? With their left feet?
- If the staircase had 191 steps, what would be the last step on which all three people would all land? Land with their left feet? Land with their right feet?

Range of Possible Teacher Questions

- What type of counting does stair skipping remind you of?
- On which stairs are Alexis and Hamilton both using the same foot?
- On which stairs are Hamilton and Dallas both using the same foot?
- On which stairs are Alexis and Dallas both using the same foot?
- Are there any strategies you can use to help you determine relationships?

Where's the Math?

This problem has interesting mathematical issues for a variety of grade levels. While searching for solutions, children may make an organized list to examine patterns. They could also use arithmetic to extend the patterns. To find the solution, the children might see a skip-counting pattern of twelve where the left feet land since 12 is the least common multiple of 2, 3, and 4.

What can we learn from the work of these students? We see that students do need some guidance in understanding a problem and selecting an approach that will give them information to arrive at a solution. Although students look for patterns while they are working, they may not make use of the pattern in the same manner as an adult. Once a problem has been explored, children are able to work on extensions that allow them to use their new knowledge.

SAMPLES OF HOW THE PROBLEM WAS SOLVED

Zach Ross and Casey Rumery submitted a summary of fifth-grade students' work that gives some interesting perspectives on how children approached this problem. The teachers presented the problem as it was written, with additional instructions for the students to record their thinking and planning.

Ross and Rumery reported the following:

> We had the students work independently for about 35 minutes. They then were allowed to work with the student sitting next to them for 20 minutes. Many questions arose while the students were working. Many students wanted clarification on questions 1, 2, and 3. They wanted to know if they were to answer the questions first or where to start.
>
> The students did begin to dive into the problem right away. The majority of the class began by drawing a picture of some sort. Fifteen of the 24 students began by drawing all 191 steps.... Many were also concerned if they would eventually find out the answer or if there was even a "correct" answer.... There were two students who sat for quite a bit of time stumped. We prompted these two students by asking them that if there were only 9 steps, could they solve the problem? They responded by saying, "probably," and they then began to get started on the problem.

Understanding the problem and selecting a strategy are both important components in problem solving. This description of student work indicates that teachers need to be aware of how to handle those instances when students are stymied in any part of the problem-solving process. Teachers should also be prepared to provide the appropriate guidance, often in the form of additional questions, to help students make progress. Teachers should encourage students to think about how they might solve the problem without drawing all 191 steps.

Ross and Rumery continued, as follows:

> Once students began working together, the ideas really began to flow. Students began to realize that "there has to be a pattern." After 45 minutes, 3 students discovered that all three of the story characters could not step on the same step with their right foot. The problem began to come along once students made a legend or code for the people and their feet. Students then compiled their information in many different ways. [See fig. 1.] Some students used grids to organize their information while others drew pictures to represent the steps. There were students who also used a variety of methods in a trial-and-error manner to figure out the answers. Many of the students acknowledged which methods did not work.

In this class, the students were encouraged to reflect on the strategies that they chose and the effectiveness of those strategies. As shown in figure

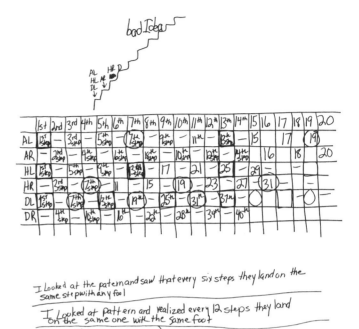

Figure 28.1. A grid was used to help solve the problem.

28.1, the students did find solutions to the questions. They were able to determine on which step from the bottom the right feet would first land. This question was asked to give some guidance to the students, but as Ross and Rumery indicated, many did not initially use this significant hint to continue the problem-solving process. We must remember that even when we try to provide a structure for students to use, they cannot use it until the problem or structure belongs to them.

Students were also able to see that all three characters would never put their right feet on the same step. Gregory supplied a "proof" of this conclusion when he wrote, "Alex will hit every step. All of A's right steps are even and H's right steps are odd." The students also discovered that although all three will land on steps 1, 7, 13, 19, 25, ... , they will land on steps 1, 13, 25, 37, ... only with their left feet. Not surprisingly, the students used a recursive idea when they reported this pattern as "every twelve steps." The students concluded that the last step on which all three would land would be step 187, and the last step on which all three would land with their left feet was step 181.

Even though the students were allowed to use a calculator, they did not seem to use the idea of counting by 12 to answer these last two questions. That is, the students did not realize that the sequences 1, 7, 13, 19, 25, ... and 1, 13, 25, 36, ... are, in some sense, "one more than" the sequences 0, 6, 12, 18, ... and 0, 12, 24, 36, ... If they had looked at multiples of 6 or 12 that are less than 191, they could have determined steps 181 and 187 as the solutions. The students' approach reminds us that the method with which we begin to solve a problem is often the one that we use throughout unless we find some reason to contemplate an alternative. At this point, a teacher might want to ask how finding the solution to this problem would help the students find a solution if the problem involved 400 steps.

Ross and Rumery did not let this problem rest:

> After the papers were collected and a few days passed, we asked the students to figure out a solution that would get the three characters up the stairs without all three stepping on the same stair. Students were allowed to work with a partner, and most of the class discovered a solution within ten minutes. The class found two solutions. In the first solution, no stair is stepped on twice. This method has Alexis starting at stair #1 and skipping 2 stairs, Hamilton starting at stair #2 and skipping 2 stairs, Dallas starting at stair #3 and skipping 2 stairs [see fig. 28.2]. In the second solution, no more than two people step on the same stair. In this solution, Alexis starts on stair #1 and skips 1 step and Hamilton starts on stair #2 and skips 1 step. Dallas can then use whatever method he would like, and only two characters step on the same stair twice.

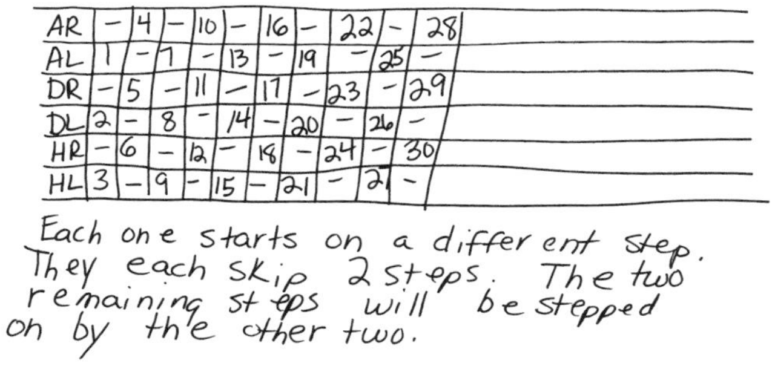

Figure 28.2. A grid was also used in a revision of the original problem.

The fact that the students needed only ten minutes to solve this extension is quite exciting and would seem to indicate that they understood the problem well and were able to make efficient use of a strategy to arrive at a solution. Both the solutions provided show remarkable insight into the structure of the problem and reflect interesting thinking by the students.

Where's the Math?

What can we learn from the work of these students? We see that students do need some guidance with understanding a problem and selecting an approach that will give them information to arrive at a solution. Although students look for patterns while they are working, they may not make use of the pattern in the same manner as an adult. Once a problem has been explored and discussed, children are able to work on extensions that allow them to use their new knowledge.

29
THREE-WAY SHARING

The Problem

Danny, Connie, and Jane have eight cookies to share among themselves. They decide that they each do not need to receive the same number of cookies, but each person should receive at least one cookie. If the children do not break any of the cookies, in how many different ways can they share the cookies?

Range of Possible Teacher Questions

- How many cookies are there now? (To determine who is conserving number)
- Can we figure this out if they don't all receive the same amount?
- How can you organize your results?
- Are there any strategies you can use to help you determine relationships?

Where's the Math?

This problem begins with conservation of number. Children need to know that "eightness" is maintained even when the eight cookies are distributed to three different people. The problem also encourages students to make a list and to be careful in recording their solutions. Creating systematic methods for counting is important in development of probability concepts.

SAMPLES OF HOW THE PROBLEM WAS SOLVED

The responses shared by teachers include student work at grades 1, 3, 4, and 5. This range indicates that this problem is accessible for students on many grade levels, although the depth of student work varies.

Both Chree Perkins, a first-grade teacher, and Mary Kay Varley, a third-grade teacher, made a connection between this problem and *The Doorbell Rang*, by Pat Hutchins. Both Varley and Perkins used the book to discuss the similarities and differences between the problem posed in the book and the "Three-Way Sharing" problem. Varley noted, "In the book, there are twelve cookies to share, but the number of children sharing them kept changing. The cookies also had to be equally shared...." Perkins wrote that after reading the book and introducing the problem, "We discussed whether it would be possible to share 8 cookies equally without breaking any if there were only 3 children. They all agreed that it would work with 4 children or with 8, but not with 3."

When Susan Vohrer presented the problem to her fourth-grade class, the students initially wondered, "Can you do two of one number?" and "Can you use turnaround facts?" Vohrer wrote, however, that they quickly "... became very absorbed in discussing among themselves different ways of coming up with an answer." Vohrer had the students discuss strategies, then put out multicolored counters, which the students immediately began using for models to help them solve the problem. Vohrer noted, "The students completely understood the idea of not having an equal number of cookies per child."

Most of Varley's and Perkins's and some of Vohrer's students made a chart or list with headings, such as D, C, and J. Some of Vohrer's students did not need such headings but were able to adequately deal with number combinations, such as 4-2-2, to signify four cookies for Danny, two for Connie, and two for Jane. Interestingly, first-, third-, and fourth-grade students all satisfactorily dealt with the idea that a 6-1-1 sharing situation would be different from a 1-1-6 sharing situation. Vohrer wrote, "This piece of information was critical and that was a definite 'Ah-ha' moment." Furthermore, in these same classes, the students determined that when the three addends were different, the result would be six different possibilities. Vohrer's students used this information in an organized fashion when they made their list of possibilities (see fig. 29.1).

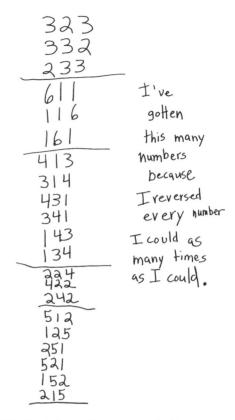

Figure 29.1. The information is presented in an organized fashion.

One student gave an excellent explanation of the process that depended on whether the possibility was a double or single numeral (see fig. 29.2). This explanation demonstrates how well some students can generalize when they are trying to solve a problem. As might be expected, some students overgeneralized, thinking that any three numbers would yield six possibilities.

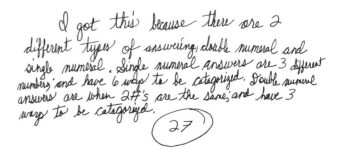

Figure 29.2. One student describes his problem-solving process.

Varley worked the problem along with her students, then shared her work with the class. Reflecting on this experience, she said, "Not one student had organized their combinations the way I had chosen to do it. When I shared my method of starting with only one child and slowly removing one cookie to determine the different ways of sharing, the students expressed amazement at the counting pattern that emerged." One of Tracy Ajello's students started trying to solve the problem in a similar manner but made an inappropriate generalization and did not obtain a correct solution (see fig. 29.3).

Two of Sandra Kelly's first-grade students organized their work in a manner similar to what Varley demonstrated to her students (see fig. 29.4). Kelly reported the following:

I am the Schoolwide Enrichment Teacher/Coordinator. I met with these students twice for about 15 minutes each time. They worked on the problem and developed their own methods of solving the problem. I provided the math journals with the problem typed into it, to record their data and solutions.... As they explained their solutions, I recorded their statements. The students found the problem to be challenging, and they were very excited about their solutions.

In discussing the work that the children did, Vohrer made interesting observations:

I was very impressed at the "math talk" that occurred in class that day. One of the best conversations was the discussion of why zero could not be used. Strategies were being shared, explanations being given, and the class was very engaged the entire time.... I felt that as I walked around the room, I could formatively assess the number sense that the different students had based on their discussions and the written words on their papers. It also helped me see the processes that the students used to arrive at their final answer.

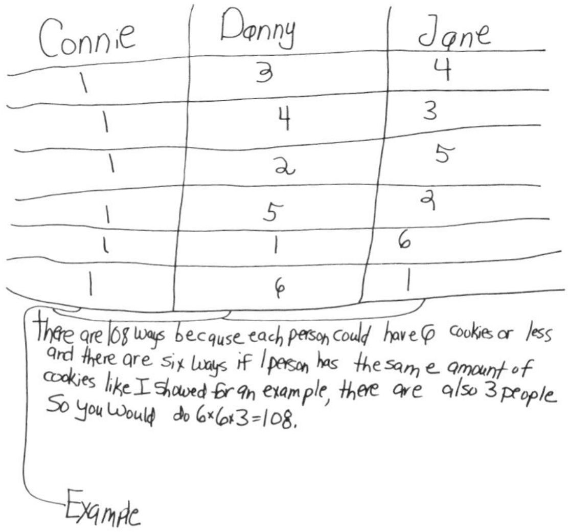

Kelly also commented on the types of problems presented in "Problem Solvers." Her school includes grades pre-K–2, uses heterogeneous grouping, adheres to a total-inclusion philosophy, and has a schoolwide enrichment model strongly in place. Hence, staff members continually look for ways to differentiate instruction. She thinks that "the mathematics problems in *Teaching Children Mathematics* are the types of activities that can be used successfully in small math groups to differentiate/enrich math lessons."

Figure 29.3. An inappropriate generalization results in an incorrect solution.

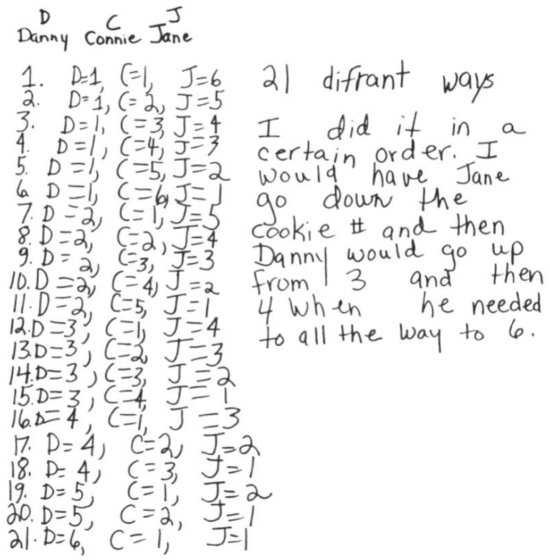

Figure 29.4. How students organized their work after the teacher demonstrated her process.

What Are the Students Telling Us?

What can we learn from the work of the students? We find that a problem can be investigated at many grade and developmental levels. Even children in first grade can understand and work with the problem and note that different arrangements are possible for the same three numbers. Problems like this one can serve as assessment opportunities for teachers and help students reflect on the strategies that they use to solve problems. These tasks also provide opportunities for both written and oral communication.